Springer Tracts in Advanced Robotics

Volume 32

Editors: Bruno Siciliano · Oussama Khatib · Frans Groen

Dejan Lj. Milutinović and Pedro U. Lima

Cells and Robots

Modeling and Control of Large-Size Agent Populations

 Springer

Professor Bruno Siciliano, Dipartimento di Informatica e Sistemistica, Universitá di Napoli Federico II, Via Claudio 21, 80125 Napoli, Italy, E-mail: siciliano@unina.it

Professor Oussama Khatib, Robotics Laboratory, Department of Computer Science, Stanford University, Stanford, CA 94305-9010, USA, E-mail: khatib@cs.stanford.edu

Professor Frans Groen, Department of Computer Science, Universiteit vanAmsterdam, Kruislaan 403, 1098 SJ Amsterdam, The Netherlands, E-mail: groen@science.uva.nl

Authors

Dejan Lj. Milutinović
Duke University Laboratory of
Computational Immunology
Hock Plaza, Suite G06
2424 Erwin Road,
Durham, NC 27705, USA
E-mail: dejan.m@duke.edu

Pedro U. Lima
Institute for Systems and Robotics
Instituto Superior Técnico
Av. Rovisco Pais
1049-001 Lisbon
Portugal
E-mail: pal@isr.ist.utl.pt

ISSN print edition: 1610-7438
ISSN electronic edition: 1610-742X

ISBN 978-3-642-09115-5   e-ISBN 978-3-540-71982-3

Springer is a part of Springer Science+Business Media
springer.com
© Springer-Verlag Berlin Heidelberg 2007
Softcover reprint of the hardcover 1st edition 2007

Data-conversion and production: SPS, Chennai, India

STAR (Springer Tracts in Advanced Robotics) has been promoted under the auspices of EURON (European Robotics Research Network) 

To my mother Dušanka and in memory of my father Ljubomir
D.M.

To my parents, Arlette and Manuel
P.L.

Foreword

At the dawn of the new millennium, robotics is undergoing a major transformation in scope and dimension. From a largely dominant industrial focus, robotics is rapidly expanding into the challenges of unstructured environments. Interacting with, assisting, serving, and exploring with humans, the emerging robots will increasingly touch people and their lives.

The goal of the new series of *Springer Tracts in Advanced Robotics (STAR)* is to bring, in a timely fashion, the latest advances and developments in robotics on the basis of their significance and quality. It is our hope that the wider dissemination of research developments will stimulate more exchanges and collaborations among the research community and contribute to further advancement of this rapidly growing field.

The monograph written by Dejan Milutinović and Pedro Lima presents an original theoretical framework in which both biological cell populations and large-size mobile robot teams can be modeled and analyzed. This unique interdisciplinary work has explored methods for discrete-event and time-continuous systems, leading to useful results for modelling and control of robot swarms. As such, the book has a wide interest for scholars in the area of multi-agent systems, including computational biology, biomedical engineering, nanorobotics and even aerospace engineering.

Remarkably, the doctoral thesis at the basis of this monograph was a finalist for the Fifth EURON Georges Giralt PhD Award devoted to the best PhD thesis in Robotics in Europe. A very fine addition to the series!

<table>
<tr><td>Naples, Italy</td><td style="text-align:right">Bruno Siciliano</td></tr>
<tr><td>March 2007</td><td style="text-align:right">STAR Editor</td></tr>
</table>

Preface

The text that is lying in front of you is based on my PhD thesis which I wrote in Lisbon between 2000-2004 AD. I call it simply "my thesis", or sometimes "my Lisbon story"[1] instead of using the long original title "Stochastic Model of Micro-Agent Populations". In recent years, I have had several opportunities to give talks about this work. Due to different time constraints on these talks, I had to tell the story in ninety, thirty or fifteen minutes. However, if I were to give a five-second talk, it would come down to these three words "Cells and Robots", the most appropriate title for this book.

In "Cells and Robots" we are presenting a theoretical framework in which both biological cell populations and large-size mobile robot teams can be modeled and analyzed. We assume that behavior of individual agents can be described by hybrid automata and we try to come up with mathematical objects appropriate for the description of the population state and its evolution. This makes our effort to study multi-agent systems different from computational models attempting to model individual agents in detail. The significance of our modeling framework for real agent populations is similar to the significance of thermodynamic laws for ideal gases in Physics of real gases.

Using the presented theoretical development, we are able to give an alternative and deeper insight into the expression dynamics of cell surface receptors. We also provide one solution to the problem of controlling large-size robotic populations.

Although we see the true value of the book only in its wholeness, a reader who is more concerned with biological applications can skip chapter 7. Similarly, a reader who is more interested in Robotics can skip chapter 5. The reader needs to have a prior knowledge of ordinary and partial differential equations and a basic knowledge of the theory of probability and stochastic processes. The reader should also be familiar with the numerical solution of partial differential equations. Knowledge of optimization and optimal control theory is helpful for the understanding of chapter 7.

The book is interesting for researchers and postgraduate students in the area of Multi-Agent systems, including natural and artificial agents, such as cells and

---

[1] After the name of the movie "Lisbon Story" by Wim Wenders, 1994.

robots, respectively. This includes researchers working in the area of the immune system modeling, Computational Biology, Bio-Medical Engineering, Robotics, Nano-Robotics, molecular devices, nano-technology and aero-space applications.

The monograph includes results related to stochastic hybrid systems and the application of optimal control for partial differential equations. Therefore, it is of certain interest to engineers and mathematicians working in the theoretical development of control theory and its applications.

My hope is that this book will inspire readers in their own research.

Acknowledgments

I wrote this book with Prof. Pedro Lima, who not only supervised me and helped me during the time I was working on my PhD thesis, but has also given me continuous support in recent years. I am grateful for his strong belief that our results have simultaneous relevance for Robotics and Biology. In recent years, we shared some difficult time trying to publish our papers resulting from my thesis. After two-three years spent in rewriting, we finally succeeded in reaching our goals. Then, to our great surprise, my thesis was also selected as one of the best PhD thesis in European Robotics.

I would like to thank Prof. Michael Athans, Emeritus Professor at MIT and currently with the Institute for Systems and Robotics in Lisbon, because I am aware that without his help my thesis, and the same goes for this book, would never have been written. It came as a result of the small project we had started in my spare time, which set up my mind to interesting research topics. I am grateful for his scientific lectures, including the one in meaning of good research. He also introduced me to the people interested in my results in their early phase and gave me the strong motivation to work on difficult problems.

During the writing of my thesis, I got exceptional help from Prof. Michael Athans and Dr. Jorge Carneiro, from Gulbenkian Science Foundation in Lisbon, who played an active role in the revising of the thesis text and their contribution is included in this book in many places. Prof. Michael Athans' major contributions are in chapters 2, 5, 6 and partly in chapter 7, while Dr Jorge Carneiro contributed to chapters 2 and 5.

Thanks also to Dr Jorge Carneiro for all of his biological lessons and interesting discussions after which I always had something new to do, including forgetting what I had previously done because it did not have a biological meaning. He motivated me to reconsider my mathematical knowledge until we reached the point of having the results meaningful to biologists. When working with him, I always felt like I spent time with a friend rather than worked hard.

Specially, I would like to thank Prof. João Sentieiro for the invitation to come to Lisbon and start my PhD studies, as well as for his friendly support and help on the way.

I would not like to forget Željko Djurović for his precious friendship from the day I came to Lisbon and Prof. Naim Afgan for his company and reading of the first draft of my thesis. My warm appreciation goes to Lionel Lapierre and Didik Soetanto for all after-lunch discussions about science and on general topics. Parts of these discussions have found their place in the text of my thesis.

Another special 'thank you' to my colleagues, Sónia Marques, Paulo Alvito, João Frazão, and Rodrigo Ventura. I was really happy working with them, sharing the office, the experience of learning and living in a great city called Lisbon.

My appreciation must go to Prof. Bruno Siciliano for his kind invitation to publish this book and his help along the way.

Last, but not least, I would like to express my warmest filial feelings for my parents, Dušanka and Ljubomir Milutinović, for giving me their constant support and help, without imposing any goals on my work and education. Unfortunately, during my PhD studies my father died and will never be able to witness the publishing of my work. Thanks to my brother Milan Milutinović for his unselfish help during all these years spent around the world. And huge thanks to my wife Dragana Milutinović for her understanding, patience, emotional support and the book proofreading.

This research was supported by the Institute for System and Robotics, Instituto Superior Técnico, Lisbon and grant SFRH/BD/2960/2000 from Fundação para Ciência e a Tecnologia (FCT), Portugal, as well as partially supported by ISR/IST pluriannual funding from FCT, through the POS_Conhecimento Program that includes FEDER funds.

Durham, North Carolina, USA   *Dejan Milutinović*
February 2007

Preface

This book is the nice result of a very fruitful multi-disciplinar work which involved researchers from different areas and, sometimes, considerably different approaches to similar problems. I use to tell my students that the benefits of such an approach clearly outstand the possible drawbacks, such as the time taken to explain to others concepts which are more or less straightforward for experts in our area, and this work is a clear example of such a statement.

Dejan's thesis was also a good example of a typical PhD student work path. When he started working with me, I gave him a topic related to robotic task modeling by discrete event systems. He actually did some work in that direction and we even published a paper about it, but finally he felt he should get involved in something that would provide a clear contribution for humankind, e.g., on health-related issues. And so he did: motivated by some class work done with Michael Athans and under the advice of Jorge Carneiro, he explored the immune system world and managed to excite me about it as well. After the thesis and the antithesis, the synthesis finally came when, one day, discussing his work, we thought that exploring a mix of discrete event and continuous state space time-driven modeling might be the right way to go to model cell population dynamics. We even went back to Robotics, when we finally concluded that what Dejan had developed could definitely be applied to the modeling and control of robotic swarms!

Overall, it was a great experience to supervise Dejan and his enthusiasm about his work. I hope this book will encourage others to pursue this research line, and act as a showcase of our friendship.

Lisboa, Portugal
February 2007

Pedro U. Lima

Contents

1. Introduction

Understanding development and functions of living organisms continuously occupies the attention of science. Consequently, mathematical modeling of biological systems is a recurrent topic in research. Recently, this field has become even more attractive due to technological improvements on data acquisition that provide researchers a further insight into such systems. Technology to read DNA sequences, or to observe protein structures along with a variety of microscopy methods, has enabled collecting large amounts of data around and inside the cell, classically considered as the smallest chunk of life.

In this book, we are particularly concerned with cells that have an active role protecting living organisms from infections caused by foreign bodies, constituting the immune system. The role of the immune system is to continuously monitor the organism, to recognize an invader, to generate a response that will clear the invader and to help healing the damaged tissues. The major components of this chain of action are motile cells. Cells motility[1] is a property intrinsic to their function, i.e., to fight against infections in the right place at the right time during an immune response. While we are quite certain about the places where cells are produced and where they reside during their life cycle, the question of how they modulate their motion and bio-chemical activity against external stimuli still presents an active field of research.

Apparently, the immune system cells are autonomous agents. On the other hand, an autonomous mobile robot can be seen as the most natural mechanic analogy of the cell. Hence, we find studies about the cell bio-chemical signal processing, intercellular communication and cell reactive behavior in close relation to signal processing, communication and control for mobile robots. Investigation of the cell-robot analogy has the potential to influence future biological research, but also to provide the guidelines for the development of a system-based approach to describe and analyze complex multi-agent/robot systems.

In the theoretical development presented in this book, we investigate one aspect of the cell-robot analogy, providing a novel approach to the study of large-size agent populations. Our approach points at understanding how individual agent dynamical behavior propagates to the population dynamics. The presented concepts and mathematical tools are general enough and provide results

---

[1] Motility - ability to move spontaneously and independently [22].

D. Milutinovic and P. Lima: Cells & Robots, STAR 32, pp. 1–8, 2007.
springerlink.com

of potential interest for *multi-agent system* applications. Multi-Agent systems (MAS), dealing with either virtual or real (robotic) agent populations, are currently a subject of major interest in engineering and computer science literature. Consequently, biological experiments involving cell populations, that are commonly used to understand cell properties, can be considered as test beds for the control strategy development for large-size agent populations. Conversely, tests that would not be possible on real organisms, can be made on robots whose behavior is designed to approximate biological models of biological populations behavior.

1.1 Analogy Between an Individual Robot and a Cell

To illustrate the analogy between an autonomous mobile robot and a motile immune system cell, we will consider a robot which is able to avoid obstacles while moving towards a goal location, and a cell which is able to move in the direction of a chemokine source. The basis for consideration of this analogy is the similarity between the ultrasonic sensor system of a mobile robot and cell receptors expressed on the cell surface.

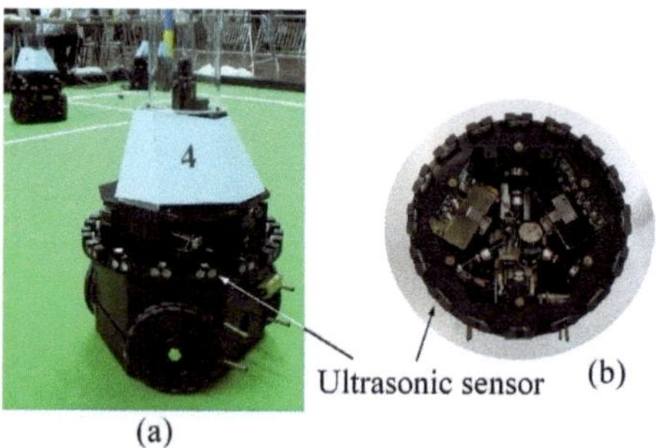

Fig. 1.1. Mobile robot endowed with ultrasonic sensors and an omnidirectional vision system. a) The robot b) Birds-eye view inside of the robot. (Institute for Systems and Robotics, Lisbon).

Figure 1.1 shows the robot equipped with a ring array of ultrasonic sensors. This system measures the proximity of the robot to obstacles in all directions. Due to its robustness and simplicity, this sensor system is usually installed on mobile robots to supplement more sophisticated vision systems.

Each ultrasonic sensor of the array is installed on the robot surface (see Figure 1.1) and the robot on-board computer reads distance measurements from each ultrasonic sensor. For each sensor in the array, the robot keeps the information about the direction and the measured distance to obstacles. The measurements are always positive and limited by some maximal distance. If the maximal distance is measured by a sensor, then the measurement is interpreted as "no obstacle in that direction".

The cell senses the environment using the receptors expressed on the surface (see Figure 1.2). Different type of receptors are sensitive to different kinds of bio-chemical environmental substances. Here, we speak only about the receptors with ability to sense chemokines. The chemokine is a substance that attracts the cell, which following the chemokine traces finds its way to different organs in the body. When, for example, the cell "decides" to move towards the lymph node, it expresses receptors that can sense the corresponding chemokine, showing the direction to its destination. Similarly, if the cell "wants" to leave the lymph node, then the cell expresses its receptors and, after some fumbling around, finds its way out. This explanation corresponds to the widely accepted idea that expressed receptor types correlate with the cell behavior. Therefore, the cell function is mainly recognized based on the receptors it expresses.

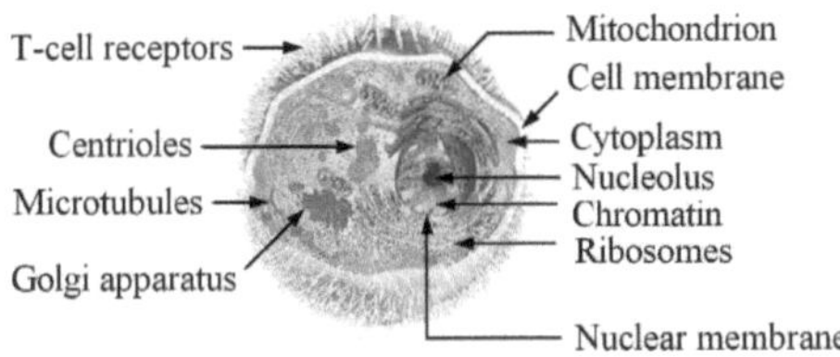

Fig. 1.2. T-cell structure: the T-cell receptors are expressed on the cell surface; the figure is based on a general cell structure [30]

The working principles of the presented robot and the cell sensory systems are quite different. However, in both cases, the sensory systems are omnidirectional, i.e., they sense the direction of the space in which the robot and the cell can move; in other words, they provide the same type of information. In the mobile robot case, the robot takes an action based on all available information, including the ultrasonic sensory system and decision making process corresponding to the robot mission task. We speculate that the cell does the similar kind of information processing before it performs an action.

This is along the idea of reverse engineering of intracellular processes. In order to understand the cellular behavior, experimental immunologists consider cells of specific phenotype and follow their behavior within a controlled scenario. Mostly they follow the cell division, cell receptors or some bio-chemical marker expression. Based on collected data, they try to understand how the cell behaves. The reverse engineering of intracellular processes directing the cell behavior is difficult, and this difficulty is easy to explain based on the analogy between the robot and the cell.

Let us imagine that someone provides us a brand new mobile robot and asks us to understand in full detail how it works without even telling us what kind of tasks the robot is able to perform. While the mechanical structure of the robot has some relevance in inferring the kind of applications which the robot is designed for, the electronic hardware structure and the relation among electrical signals controlling the robot, provide a little insight into the structure of the robot control software.

Although reverse engineering of an individual cell seems important, it ignores the fact that cells are rarely under physiological conditions isolated from other cells. To have a complete picture of the cell functionality, its communication and cooperative behavior with other cells must be considered as well. In the next section, we deal with cell population models and the similarity between robot teams and cell populations.

1.2 Robot Teams and Cell Populations

One way to disclose the analogy between a robot team and a cell population is through ordinary differential equation (ODE) models which are exploited to describe the population dynamics in both cases. Inspired by the Lotka-Voltera predator-prey equation, ODEs are used to model anti-viral immune responses where, in the simplest form, the immune system cells play the predator's role and the virus or infected cells play the prey's role. More realistic ODE models of the immune system response include different cell types playing the roles of effector cells, memory cells, helper cells, naive cells, etc. These ODE models describe the cell amounts in each specific cell type, each of them dedicated to a specific task. The cell, just like a robot, can switch from one task to another. Hence, the similarity of the immune response ODE models to models of task allocation among the robots of a robot team is not surprising. Likewise, the analogy between the cell death and the robot failure is not unexpected; in both cases the agent remains dysfunctional. The major difference between the robot team and the cell population is the result of the state-of-the-art electro-mechanical robot design not enabling a property similar to the cell division.

Figure 1.3a is a computer-generated image[2] of the lymph node, as seen by the use of two-photon microscopy [51, 52]. It shows a dendritic cell which is an antigen-presenting cell (APC) and T-cells. The APC expresses, on its surface, pieces of pathogen protein, so-called *antigen*, signaling the presence of pathogen, for example of virus or bacteria. The contact of APCs and their interaction with T-cells is of vital importance for the start of the immune system response. Without any infection, the T-cell moves randomly around the lymph node and scans APCs for the presence of the antigen. However, when an infection is present, APCs bearing antigen peptides drain to the lymph node, causing a great chance of a T-cell meeting such APCs. When this happens, the T-cell matures and starts the immune response, but this does not happen straightforwardly. Actually, it seems that the T-cell must conjugate to more APCs, before it develops a longer-lasting contact to the APC, stimulating the T-cell enough to develop an

---

[2] The image cells morphology is modeled by the point light sources convoluted with the samples of 2D Gaussian distribution. The source positions are sampled from the numerical solution of 2D stochastic differential equations.

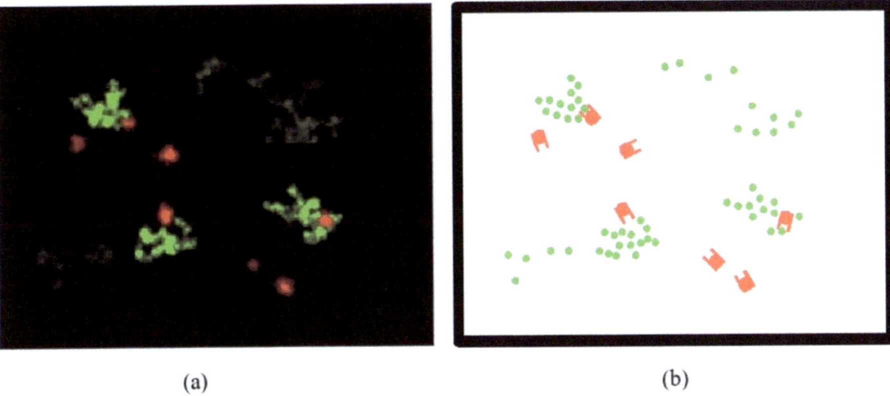

(a)   (b)

Fig. 1.3. a) Computer-generated image of a cellular interaction inside the lymph-node: T-cells (red) scan dendritic cells (green) (computer simulation by D. Milutinović) b) Robotic team (red) collecting pucks (green)

immune response. Consequently, the more the T-cells, the faster they scan the APCs inside the lymph node, and the earlier they recognize the presence of the antigen and possibly develop a strong immune response. However, while scanning, T-cells also compete for APC surface sites providing stimulating signals to them. It is also worth mentioning that sometimes T-cells are not able to get straight to the APC sites because they must avoid other T-cells.

A similar kind of scenario is studied in Robotics [44] in order to understand the cooperative behavior and task allocation of a robotic team (see Figure 1.3b). In this scenario, robots are moving inside a limited operating space. The robot team mission is to locate and collect pucks that are distributed over the operating space. To succeed in that task, they should also avoid collisions with their teammates and conflicts regarding common resources.

The main distinction between biological and robotic examples lies in the fact that cells move in three dimensions while agents in the robotic case move in a two dimensional space. In all other aspects, biological and robotic examples are analogous, since the cell division, a major difficulty in the cell-robot analogy, does not appear in the early phase of the immune response we described in this section.

1.3 Related Work

Social insect communities, such as ant and bee colonies, are among the first investigated large-scale multi-agent systems [11, 17]. They are an example of self-organized systems [59], where the colony behavior emerges from the behavior of individual insects. The main property of the insect colonies is that not only does their behavior seem intelligent, but also robust to environment changes and the colony size. This motivated the research on virtual [23, 65, 84] and robot [4, 6] agents which applies concepts found in biological systems to multi-agent systems.

For the analysis of large-size robotic populations [32], control performance [74] and formation feasibility [73], deterministic models are used. These models are related to the robotic formation, i.e., the relative positions of the robots in the operating space. On the other hand, the task allocation and the task performance of groups of robots are modeled under a probabilistic framework in [1, 10, 29, 48, 71]. The ODE models resulting from this framework are in close relation to the ODE models used in the immune system modeling. To illustrate this, we use, in the previous section, the robotic scenario from [43, 44, 49], where the same framework is applied.

The modeled level is the main source of difference among alternative immune system modeling approaches. The immune system behavior can be modeled at the level of *bio-molecular interactions, signal transduction, cell-to-cell interaction, lymphocyte population dynamics*, or the *immune response* to virus and bacteria infections. ODE models are used to model the anti-viral immune response [9, 60, 61, 82], population dynamics of lymphocytes in which the interaction is based on idiotypic networks [58, 62, 72, 80], resource competition, or more complex resource competition and suppression interaction [7, 12, 37, 41, 42].

In the case when immune system phenomena show dependence on spatial heterogeneity, partial differential equation (PDE) models are applied, such as in the modeling of immunological synapses [14, 83] or repertoire dynamics in "shape space" [67]. Wide use of ODE and PDE models is preferable because of developed and well-known mathematical properties of their solutions. However, the problem with differential equation models is a large number of (physically meaningful) parameters. Moreover, the parameter values strongly depend on the proper scaling of variables, and some of them are difficult to identify by experiments.

Assuming that the parameters of interaction are not of great importance in understanding robust immune system mechanisms, which might be correct in the case of signal transduction and cells activation, networks of interconnected automata, i.e., Boolean network models, have been introduced [36]. Using Boolean network, feedback loops and stationary shapes of those loops can be calculated. Along this line of reasoning, in the case of spatial heterogeneity, cellular automata models are used in [16, 40, 70]. In immunological synapse modeling, the space is physical space and in idiotypic networks, the space is "shape space".

The list of references relating to modeling of multi-agent systems and the immune system presented above is far from being complete, but it gives a flavor of the state-of-the-art in both fields. In this book we use a hybrid system approach [76] to modeling. This approach has been used to model intra-cellular bio-molecular interactions [3]. The model employed is a *deterministic hybrid system* model of a single interaction, and the authors conclude that studying the bio-molecular interaction using *stochastic hybrid systems* [35] is a challenging issue.

One of the major problems of modeling in immunology is to derive macroscopic properties of the system from the properties and interactions among the elementary components [62]. The hybrid system approach presented in this monograph is an approach to such problems. The motivation for studying this problem comes from the modeling of T-cell receptors (TCR) expression dynamics. The available models of TCR expression of the T-cell population interacting with APCs are ODE models [5, 69, 83].

Inspired by the analogy between a biological cell and a robot, we also find challenging to exploit the hybrid system modeling approach for a model-based controller of a large-size robotic population. As a result, we obtain the centralized controller which is based on the Pontryagin-Hamiltonian [21] optimal control theory for PDEs [46] and provides control of the population space distribution shape. The centralized optimal controller we introduce here is based on a new concept for the control of one class of stochastic hybrid automata, where the state probability density function of stochastic hybrid automaton is controlled. The controller is truly optimal and is not based on discrete approximations [35] or piecewise affine approximations [57]. However, as in classical optimal control [50], it is an open-loop controller and there is no warranty that the control can be expressed analytically. For the numerical computation of optimal control, discrete approximations in time and space are necessary.

1.4 Book Outline

The presented research has a strong multi-disciplinary character. It is motivated by the investigation of the TCR expression dynamics of a T-cell biological population and the results are extended towards a system approach to study the population composed of a large number of individual agents. A brief summary of the contents of the eight book chapters, besides the current Introduction, follows:

Chapter 2. Immune system and TCR dynamics of a T-cell population. This chapter introduces the problem which motivated this work. The basic biological facts about the TCR expression and previous modeling efforts based on ODE equations are described. We conclude that existing ODE models are unsatisfactory in relating the modeled dynamics of the average TCR expression of the T-cell population and the TCR expression dynamics of the individual T-cells. We also present the experimental setup which is used in the verification of those ODE models. Taking into account that by using only average values of measurements we throw away potentially useful information, we again arrive to the conclusion that for getting better insight into the TCR expression dynamics a different modeling approach has to be examined.

Chapter 3. Micro-Agent and Stochastic Micro-Agent Models. Given the biological facts, the hybrid systems framework appears as a natural framework for modeling of the TCR expression dynamics. In this chapter, hybrid system models of a T-cell and of a T-cell population are introduced. The model of the T-cell motivates the formal definition of a Micro-Agent. We use the Micro-Agent model of an individual T-cell to build a T-cell population model, where the T-cell dynamics of individual cells is decoupled from the complex population dynamics. By applying a stochastic approximation to this population model, a Stochastic Micro-Agent model of the population is developed.

Chapter 4. Micro Agent Population Dynamic Equations. In this chapter, we discuss the Maxwell-Boltzmann distribution to relate micro- and macro-dynamics of interactions. The idea underlying the probabilistic description of the relation between micro- and macro-dynamics is borrowed from statistical physics and is used for the development of the mathematical analysis of the agents population. The PDE describing the probability density function dynamics of a Stochastic Micro-Agent, whose discrete state can be modeled as a Markov Chain, designated as a *Continuous Time Markov Chain Micro-Agent* $(CTMC\mu A)$, is derived.

Chapter 5. Stochastic Micro-Agent Model of TCR dynamics. The application of the developed theory to study the TCR expression dynamics is presented in this chapter. First, we present a numerical example which illustrates the application of the PDE derived in Chapter 4. Using this example we discuss the qualitative difference between the results obtained by the ODE and PDE models based on the same physical assumption regarding TCR dynamics. Next, we perform the steady state analysis of the PDE which provides us with the capability to predict the TCR steady state distribution of the T-cell population. The predicted steady state is compared with experimental data. Finally, we analyze experimental data which allows us to study only the TCR expression down-regulation. Using our model, we test linear and quadratic hypotheses of the TCR down-regulation and we identify the parameters of the dynamics.

Chapter 6. Stochastic Micro-Agent Model Uncertainties. In this chapter, motivated by the problem of modeling diversity in the T-cell APC interaction, we develop a method to handle a $CTMC\mu A$ continuous dynamics parameter uncertainty. The method is developed towards the systematic modification of the original $CTMC\mu A$ to the $CTMC\mu A$ model which incorporates the parameter uncertainty. The parameter uncertainty is described by its probability density function (PDF). Discrete and continuous parameter uncertainties are discussed.

Chapter 7. Stochastic Modeling of a Large Size Robotic Population. Here we apply our approach to the modeling of a large population of robots. To motivate this development, we introduce an example of a robotic population modeled by a Stochastic Micro-Agent. For this example, we predict the evolution of the robots position probability density function. This prediction illustrates that different parameters of the stochastic behavior lead to different densities of the robotic population in probabilistic space. Based on this, we introduce the optimal control problem of controlling the shape of robots position PDF, and the application of Minimum Principle for PDEs to this problem is discussed.

Chapter 8. Conclusions and Future Work. In this chapter, we present the conclusions of this research and some possible directions for future work based on the theory and examples given in this book.

2. Immune System and T-Cell Receptor Dynamics of a T-Cell Population

Figure 2.1 presents the two parts of the immune system [33]. The *Innate immune system* is made of cells and molecules that are genetically encoded and co-evolve with pathogens[1]. It is inherited as such and it represents the first line of the organism defense. It can recognize some pathogen, such as virus or bacterium,

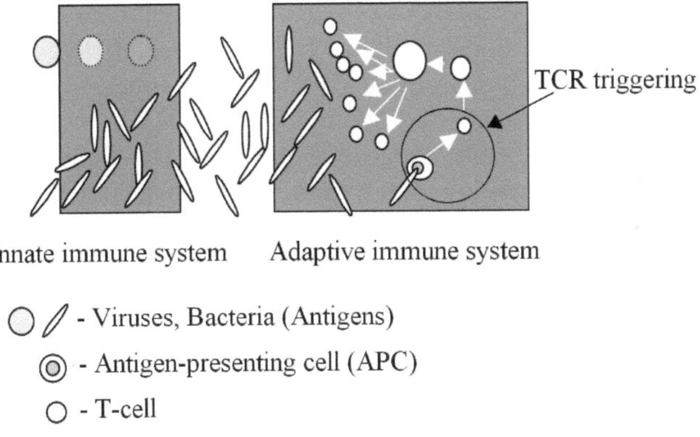

Fig. 2.1. Innate and Adaptive Immune System

and remove them within several hours. However, the diversity of pathogens it can recognize is limited and some pathogen can escape the defenses of the innate immune system. Then the pathogens are faced with the response of the *adaptive immune system*. The adaptive part is able to recognize a larger diversity of possible antigen[2] carried by pathogen. During the invasion phase the population

---

[1] Pathogen - any disease-producing agent (especially a virus or bacterium or other microorganism) [22].

[2] Antigen - any substance (as a toxin or enzyme) that stimulates an immune response in the body (especially the production of antibodies) [22]. In our context, piece of protein constituting a pathogen.

D. Milutinovic and P. Lima: Cells & Robots, STAR 32, pp. 9–14, 2007.
springerlink.com

of pathogen increases. The adaptive immune system is programmed to recognize the antigen and produce enough effector cells, specific to the pathogen, to suppress and eliminate the pathogen population. That process typically takes several days. During this time, the pathogen could be causing considerable harm, and that is why the innate immunity is so essential.

The adaptive immunity is orchestrated by different types of T-cells . The life-history of these cells is critically dependent on signals via their receptors expressed on the cell surface. T-cell receptor (TCR) signals are involved in the positive and negative selection of immature T-cells in the thymus [77], and also in the survival, activation, differentiation, and cell cycle progression of mature circulating T-cells [63, 81]. Understanding of immune responses and the dynamics of T-cell populations, therefore, requires a basic understanding of the signaling processes.

An *antigen-presenting cell* (APC) is the immune system cell which processes an antigen and transfers information about it to the T-cell. Using this information, the T-cell builds an immune response which generate the cells able to clear the pathogen. The mechanism initiating this information transfer is called TCR triggering and its place in the immune system feedback loop is illustrated by the circle in Figure 2.1. As a consequence of this "information transfer", the TCR expression level of the T-cell surface decreases. The amount of TCRs on the surface of the T-cell is a measure of the intensity of this interaction. This amount changes with time and understanding of the dynamics of this change is an important step towards understanding of the immune response dynamics. To study the TCR triggering mechanism and dynamics of the TCR down-regulation, biologists use populations of T-cells exposed to APCs. In this book, we deal with the modeling of the TCR expression level dynamics of such populations. Throughout the book we will call the TCR expression level shortly as TCR expression and its corresponding dynamics TCR expression dynamics, or just TCR dynamics.

This chapter starts by the description of a minimal biological system featuring the TCR triggering dynamics and referencing the works concerned on the modeling of TCR down-regulation dynamics of T-cell population in Section 2.1. In Section 2.2 we continue with the explanation of the experimental setup which is used for the verification of the proposed models.

2.1 Surface T-Cell Receptor Dynamics in a Mixture of Interacting Cells

T-cell receptor signals are triggered by TCR-ligands, MHC-peptide complexes, which are membrane molecules presented by specialized cells, called antigen-presenting cells (APCs), see Figure 2.2. TCR triggering requires that the T-cell and the APC form a conjugate which enables that the receptor and the ligand interact with each other. Thus, a realistic model of TCR signaling should take into account not only the dynamics of the components of the signaling cascades, but also the dynamics of the processes of APC-T-cell conjugate formation and conjugate dissociation.

Fig. 2.2. T-cell receptor triggering: T-cell, T-cell receptor (TCR), antigen-presenting cell (APC), peptide-MHC complex

The minimal biological system with the properties we are interested in is a homogeneous mixture of interacting T-cells and APCs which present TCR-ligands, depicted in Figure 2.3. In order to model biological experiments, we consider that the T-cells and APCs are moving randomly under the forces that are the result of intra-cellular or environmental conditions. Because of that, we assume that T-cells and APCs form transient conjugates. We should point that when we have started our theoretical development, this assumption was not based on firm biological evidences. Therefore, it was under question whether our theoretical work had any value in describing biological reality. However, recent *in vivo* experiments, where a two-photon microscopy has been exploited to observe the motion of T-cells inside the lymph node [51, 52], has resolved this issue. In physiological conditions of the lymph node, the T-cells move randomly and form transient conjugates with APCs.

The consequence of the TCR signaling, that can take place after the conjugate formation, is a decrease in the cell surface, i.e., the cell membrane TCR expression resulting from internalization of those membrane TCRs that have been triggered by the ligand. When the two cells dissociate, the TCR amount is slowly restored or remains constant.

In an attempt to understand the effects of the recurrent APC-T-cell interactions in the induction of self-tolerance and in the regulation of T-cell population sizes this kind of processes has been simulated [68]. Although this analysis provided some insight into these processes, the scope of conclusions was limited by the fact that no fully analytical model was derived.

Available analytical models are ODE models of the average TCR expression of the population [5]. To derive such a type of model, PDE equations [83] and a mean field approach [69] are used. However, in both models the assumption that the dynamics of the T-cell population can be explained using the concept of an *average cell* is used. This means that the population TCR average expression dynamics is considered equivalent to the dynamic responses of an *average cell*. As this assumption is unrealistic, different modeling approaches, which will properly treat relations between the TCR expression dynamics of individual T-cell and the TCR average expression of the population, must be considered.

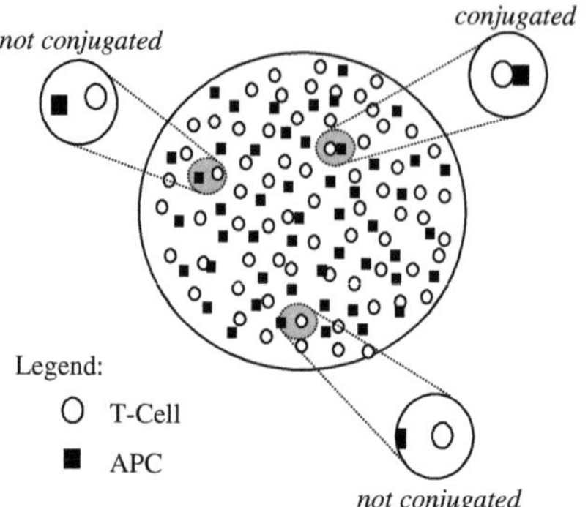

Fig. 2.3. T-cell population surrounded by the APCs: T-cells can be *conjugated* or *non conjugated* to APC [53, 54, 56], © 2003 IEEE

The models in [69] and [83] are tested against experimental data [75], where the CD3 receptors expression level is used as an indirect measure of the amount of all receptors present on the T-cell surface. The experimental setup which is used to produce experimental data analyzed in this book is described in [45, 75] and presented in the following section.

2.2   T-Cell Receptor Triggering Experimental Setup

The biological experimental setup, which is used for TCR expression dynamics recording, consists of a few identical samples of T-cell-APC population and a special device called *Flow Cytometry scanner* (FCS). The experimental setup, as well as the Flow Cytometry scanner working principle, are depicted in Figure 2.4. The Flow Cytometry scanner has a special construction which exposes one cell at a time to the laser light. For different type of analysis, different wavelengths can be applied. The intensity of scattered light, or light emitted by the cell after exposure (fluorescence), can be recorded in digital format after an analog to digital conversion. In our experimental setup, this fluorescence intensity has a particular meaning because the TCR molecules are labeled by a fluorescent probe. Therefore, the fluorescence intensity corresponds to the TCR amount expressed on the T-cell. By exposing all the cells from the sample to the scanner, the TCR expression of each cell and the average expression of the TCRs within the sample can be estimated.

Fluorescent light intensity measurements of the population can be represented in the form of the distribution where on the x axis is the light intensity and y axis is the quantity of T-cells with a given light intensity x. An example of T-cell population measurements is given in Figure 2.5.

Once the sample has been exposed to the FCS, it is destroyed and cannot be re-used for another measurement. Therefore, experimentalists make a series of identical T-cell-APC mixtures. Interaction within all the samples start at the

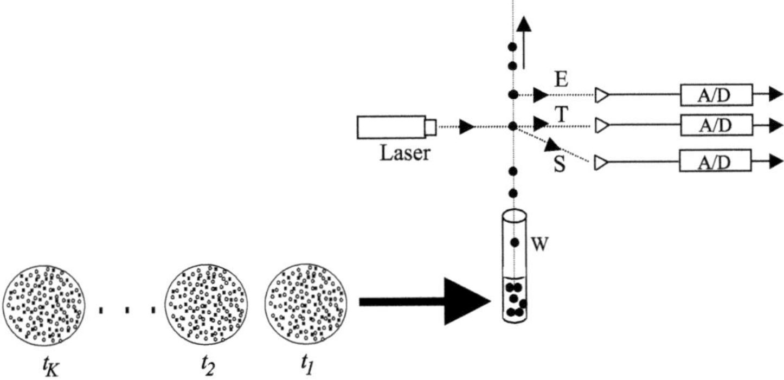

Fig. 2.4. Experimental setup and Flow Cytometry scanner working principle :
W - T-cell container, S - scattered light, T - transmitted light, E - emitted light,
A/D - analog to digital converter, t_1, t_2,...t_K- the samples which are exposed to the
scanner at time t_1, t_2,...t_K

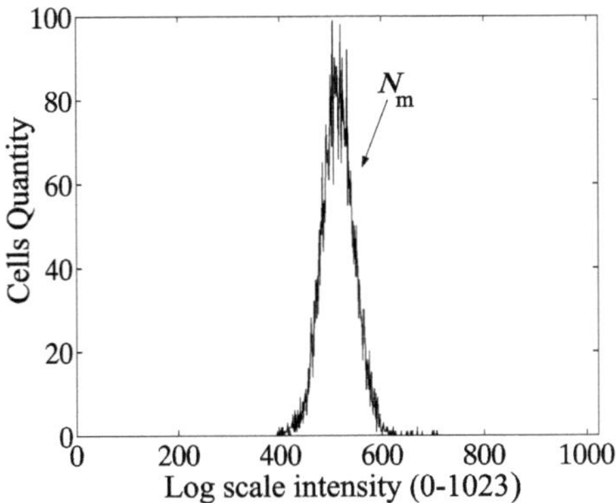

Fig. 2.5. Flow Cytometry measurements distribution of one T-cell population, by Lino
[45]; the full scale from 0-1023 covers 4 decades

same time. Under these conditions, we assume that exposing the first sample to
FCS at time t_1, the second sample at time t_2 etc., produces a sequence equivalent
to the non-destructive observation of a single sample at times t_1, t_2,...t_K. See
Figure 2.4.

Using this experiment, the average TCR expression sequence at times t_1,
t_2,...t_K is calculated and compared to the sequences predicted by the ODE
models. It is obvious that data samples used in the ODE models parameter
identification are the average of TCR expression of a few thousand cells, i.e., the
average of a few thousand measurements. Considering only the average value of
measurements means that we throw away a lot of potentially useful information

which is part of the Flow Cytometry measurement distribution of the cell population. To incorporate such information in a mathematical model, we come again to the conclusion that a different approach to the study of TCR expression dynamics should be considered.

2.3 Summary

In this chapter, the basic biological facts about TCR triggering and TCR downregulation are presented. The previous studies are based on the ODE models of the average TCR expression of the population. To verify this kind of models, the average value of a large amount of data is considered. Therefore, we find interesting to exploit these data and investigate how the biological facts about the individual dynamics of TCR expression can be used to make inference about the average TCR expression dynamics of the population.

3. Micro-Agent and Stochastic Micro-Agent Models

In this chapter, a hybrid automata approach to the modeling of individual agents and population behavior is presented. A hybrid system consists of an event-driven discrete state component and a time-driven continuous-state component [76]. Hybrid automata are particular cases of hybrid systems, where the discrete-state dynamics is modeled by a finite-state automaton. The application of a hybrid automata approach is motivated by the discussion in Section 3.1. The model of an individual agent is called Micro-Agent [53, 56]. This model is illustrated by the T-cell hybrid automaton model [54] in Section 3.2. This biologically inspired modeling problem is resumed in Section 3.3, where the Micro-Agent-based model of a T-cell population is presented. This population model includes the full population complexity. Approximating this complexity by stochastic modeling leads to a *Stochastic Micro-Agent model*. Biologically motivated Micro-Agent and Stochastic Micro-Agent are formally defined in Section 3.4 and Section 3.5, respectively.

3.1 Problem Formulation

Our problem is motivated by the search for modeling methods which will give us better insight into the relation between dynamics of individual cells and cell population dynamics. In the problem formulated below, we withdraw *the average cell* concept and propose to replace it by population macro-dynamics, which results from the propagation of the individual cell dynamics, i.e., micro-dynamics.

Interaction between T-cell and APC starts when they become conjugated due to a random encounter. Thus, considering single T-cell, the event *conjugate formation*, is a random event. Similarly, T-cell and APC are also subject to the opposite process, *dissociation*, when the interaction between T-cell and APC ends. Although we can not easily conclude whether dissociation is a random or deterministic process, we still can define event *conjugate dissociation*, which happens at the time point when the interaction between T-cell and APC ceases. Based on this discussion, it is clear that the behavior of an individual T-cell can be described by a hybrid automaton with discrete states *conjugated* and *free*. In

D. Milutinovic and P. Lima: Cells & Robots, STAR 32, pp. 15–23, 2007.
springerlink.com   © Springer-Verlag Berlin Heidelberg 2007

each discrete state the TCR expression of an individual T-cell changes and the transition between the discrete state is the result of *conjugate formation* and *conjugate dissociation* events.

The complete TCR expression dynamics of the T-cell population is composed of TCR expression dynamics of each individual T-cell and the dynamics of the cell motion, which leads to *conjugate formation* and *conjugate dissociation* events. Therefore, the TCR expression dynamics of the T-cell population is complex due to the amount of variables necessary to describe the state of the complete population. If we assume a 3D model of motion, we need 6 state variables per T-cell just to explain the position and velocity of T-cell. We need also at least one state variable, if the state of TCR expression dynamics is the amount of TCRs, for TCR expression dynamics state, and one discrete variable that contains information on whether the T-cell is conjugated or free. In total, this means, at least, 8 state variables per T-cell. A population of 1000 T-cells requires a state vector of the dimension 8000. Therefore, to make a detailed simulation of the 1000 T-cell population, we need to update, at each step, 8000 variables. However, the problem of how the individual cell TCR expression dynamics propagates to the population statistics, such as the average value or the variance of the population TCR expression, could still be answered.

The problem of matching the individual state of TCR dynamics to the population observation data is even more difficult than the problem of simulation. Taking into account only a discrete part of the state space, the population can be in about 10^{300} different states. If we would like to match some experimental data to our model, we should estimate in which of these 10^{300} states our system is, at each time instant, clearly an impossible task.

From a system theory point of view, the relation between the individual behavior of the TCR expression level, after the triggering, and its link to the observation of the experiments made with the T-cell population, leads to the following question:

> *How do the individual cell dynamics aggregate to the macro dynamics of the population and experimental measurements?*

Since the individual dynamics describes the population behavior at the micro level, the individual dynamics will be called *micro-dynamics* in the sequel. A systematic approach to study the relation between the micro- and macro-dynamics of a large population of individuals, such as biological cell populations or multi-robot populations, is a major problem investigated in this work.

3.2 T-Cell Hybrid Automaton Model

Considering the TCR expression dynamics, the state of the T-cell is composed of continuous and discrete states. The continuous state x is related to the TCR expression dynamics while the discrete state q describes whether the T-cell is conjugated to an APC, or not. Thus, hybrid automata methodology appears as a natural modeling framework for the biological T-cell population.

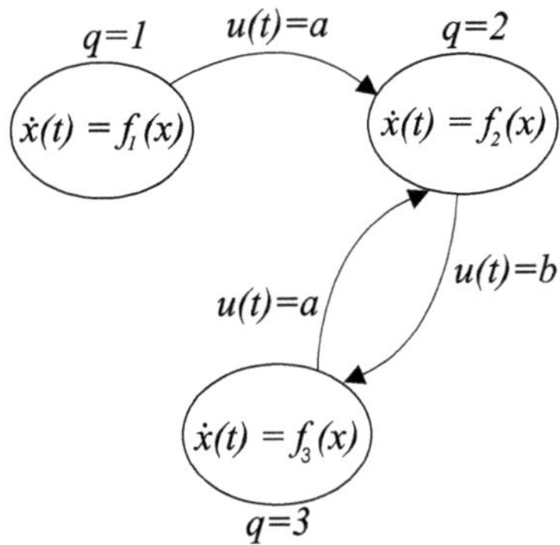

Fig. 3.1. Hybrid automaton model of the T-cell - APC interaction, $x(t)$ - TCRs expression level, $u(t)$ - event sequence, discrete states q: 1 - *never conjugated*, 2 - *conjugated*, 3 - *free*; events: a - *conjugate formation*, b - *conjugate dissociation*; $f_q(x)$ - the TCR expression dynamics of discrete state q, $q = 1, 2, 3$ [54]

The hybrid automaton model of the T-cell is presented in Figure 3.1. By this model, the T-cell can be in one of three discrete states: *never conjugated*, *conjugated* and *free*. This is a consequence of T-cell expression dynamical behavior we expect in this biological system. The state *never conjugated* means that T-cell has never been conjugated to any APC. The state *conjugated* means that the T-cell is currently conjugated to an APC and *free* means that T-cell has been conjugated before with some APC, but then becomes free. Transitions among the discrete states are consequence of the T-cell and APC motion dynamics. In this modeling approach, we assume that motion dynamics produces the time sequence of the events $u(t)$ which changes the discrete state of the T-cell. This time event sequence is defined at each time instant and takes value a, b or ε. The symbol a and b stands for *conjugate formation* and *conjugate dissociation* event respectively. Symbol ε is introduced to describe *no event*, which means that the discrete state is not changing.

The discrete states are introduced to model different TCR expression variations of the T-cell, when the T-cell is conjugated to APC or free. The TCRs expression dynamics of each discrete state is assumed to obey an ODE of the type

$$\dot{x}(t) = f_q(x), \quad q = 1, 2, 3 \tag{3.1}$$

where x is the expression level of TCRs and $f_q(x)$ defines the TCR dynamics in each discrete state, $q = 1, 2, 3$,i.e., *never conjugated*, *conjugated* and *free* discrete state, respectively. The time event sequence $u(t)$ is a mapping

$$u(t): \ R \ \rightarrow \{a \ , b \ , \ \varepsilon\}, \ t \in R \ (t \text{ is a real number representing time}) \tag{3.2}$$

At this modeling level, it is not important what the functions f_q are and whether the sequence $u(t)$ is deterministic or stochastic. However, it is important to note

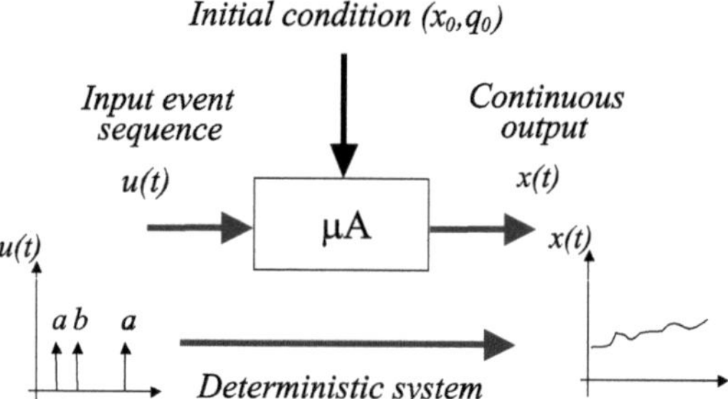

Fig. 3.2. Micro-Agent model of the T-cell ; $u(t)$ - event sequence; events: a - *conjugate formation*, b - *conjugate dissociation*; x - expression level of TCRs; x_0 - initial expression level of TCRs, q_0 - initial discrete state [54]

that the T-cell hybrid system model (Figure 3.1) is a deterministic model. Given an initial state (x_0, q_0) and a time sequence $u(t)$, the expression level of TCRs $x(t)$ and the discrete state $q(t)$ can be calculated in a deterministic way.

Taking into account its deterministic nature, each T-cell model can be represented as a deterministic *single-input, singe-output* (SISO) system, as in Figure 3.2. The input to this system is a sequence of the events $u(t)$ and the output is the expression level of TCRs $x(t)$, which is a continuous time function. This $input - output$ representation will be designated as a Micro-Agent (μA) model of the T-cell .

3.3   T-Cell Population Hybrid System Model

The complete population model can be derived by modeling each T-cell of the population by a T-cell Micro-Agent model. However, simple collection of the Micro-Agent models will not reflect the TCR dynamic of the biological population. The part of biological population complex dynamics which is not covered by this collection reflects the complex dynamics of the population producing the conjugate formation and conjugate dissociation of T-cells and APC. This complex dynamics makes the difference between the TCR dynamics of a collection of separate T-cells and the TCR dynamics of a T-cell biological population (Figure 3.3).

The *conjugate formation* and *conjugate dissociation* events, of T-cells in the population, are result of a complex dynamics which depends on the amount of the cells, their position, speed, orientation, geometry, etc. In the model of a T-cell population we are proposing, this complex dynamics is represented by the Population Event Generator (PEG) block in Figure 3.3 . The PEG has as many outputs as Micro-Agents (T-cells) and generates the events sequences $(u_1, u_2, \ldots u_N$, Figure 3.3) exactly in the same way as they appear in the biological population. This is graphically depicted in Figure 3.3 by the arrows pointing from the cells in the population to the Micro-Agents input. The introduction of

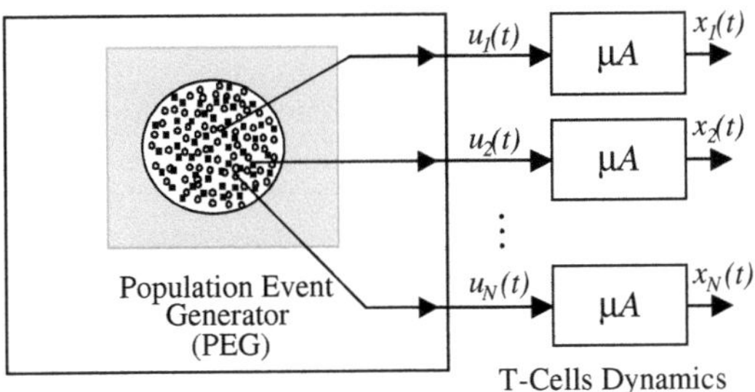

Fig. 3.3. T-cell population hybrid system model, u_i - event sequence input to the ith T-cell, x_i - the TCR expression level of the ith T-cell, $i = 1, 2 \ldots, N$ [54]

the PEG in the model makes a strong practical point in the decomposing of the TCR population TCR dynamics into [54]:

- a deterministic part (Micro-Agents), which describes the behavior of the individual T-cell TCR expression dynamics,
- a stochastic part, related to the complex dynamics of massive random encounters of T-cells and APCs (PEG).

The event generation has a complex dynamic process which depends on many variables in the population. An approach to model this complexity is to apply a stochastic approach, *where the full complexity of the interactions is described by the probability that an event happens* [54]. The population we are considering consists of individuals of the same nature and we expect that the event sequences $u_i(t)$, $i = 1, 2, \ldots N$, generated by the PEG, are of the same stochastic nature as well.

Under the assumption that the event sequences are mutually independent, the PEG can be decomposed into the set of parallel Micro-Agent Event Generators

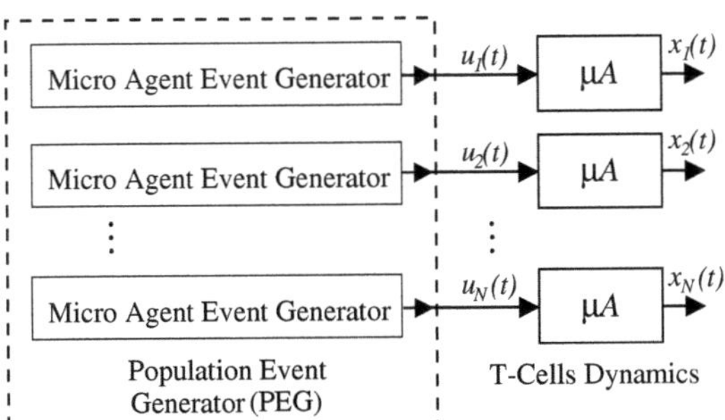

Fig. 3.4. T-cell population hybrid system model, where the Population Event Generator (PEG) is decomposed into several Micro-Agent Event Generators (MAEG). The serial connection of MAEG and Micro-Agents is named a "Stochastic Micro-Agent" ($S\mu A$); u_i - event sequence input to the ith T-cell, x_i - the TCR expression level of the ith T-cell, $i = 1, 2 \ldots N$ [54].

(MAEG), as it is presented in Figure 3.4. Each of the MAEGs produces an event sequence to the input of one Micro-Agent. We name the serial connection of MAEG and Micro-Agent a "Stochastic Micro-Agent". The output of the Stochastic Micro-Agent is a continuous time stochastic process.

The hybrid-system model of the biological population we are proposing is a collection of Stochastic Micro-Agents, Figure 3.4. In the following section we will introduce the formal definition of the Micro-Agent, μA, and Stochastic Micro-Agent, $S\mu A$.

3.4 Micro-Agent Individual Model

The aim of this section is to introduce a formal mathematical definition of the population building block, denoted as Micro-Agent. The Micro-Agent is a hybrid automaton [53, 54, 56]. Therefore, the definition of a hybrid automaton is introduced first. The abstract definition of a Micro-Agent is necessary in order to develop the system theory approach of this monograph.

Definition 1 [76]. A hybrid automata H is a collection *H=(Q, X, Init, f, Inv, E, G, R)*, where:

- Q is a finite set of discrete states,
- $X \subseteq R^n$ the continuous state space,
- $Init \subseteq Q \times X$ is the set of initial states,
- $f : Q \times X \to TX$ assigns to each $q \in Q$ a vector field $f(x, q)$,
- $Inv : Q \to 2^X$ assigns to each $q \in Q$ an *invariant* set; as long as the discrete state is $q \in Q$, then the continuous state $x \in Inv(q)$,
- $E \subseteq Q \times Q$ is a collection of edges (discrete transitions),
- $G : E \to 2^X$ assigns to $e \in E$ a *guard* set, representing the collection of the discrete transitions allowed by the state vector,
- $R_s : X \times E \to X$ assigns to $e \in E$ and $x \in X$ a reset map, describing jumps in the continuous state space due to event e.

Definition 2 [53, 54, 56]. A Micro-Agent μA is a single-input multi-output hybrid automaton, defined as a collection $\mu A = (H, U, \tau, Y)$, where:

- H is a hybrid automaton $H = (Q, X, Init, f, Inv, E, G, R)$ that satisfies the following properties:
    - $X = R^n$, the state space of the continuous piece of H,
    - $Inv(q) = X, \forall q \in Q$, i.e., for any discrete state $q \in Q$, the invariant is the full continuous state space,
    - $G(e) = X, \forall e \in E$, i.e., all defined transitions are allowed,
    - $R_s(e, x) = x, \ \forall (e \in E \wedge x \in X)$, i.e., the transition e does not change the continuous state x,
- U is a finite set of input discrete events, including the *nil* event ε,

- $\tau : U \times Q \to E$ assigns to the pair formed by the discrete event $u \in U$ and discrete state $q \in Q$ the transition $e = (q, q') \in E$, where $\tau(\varepsilon, q) = (q, q)$,
- $Y = R^m$ is the output state, a μA output $y \in Y$ is a function of the continuous state x, $y = g(x)$.

Remark 1. The Micro-Agent state is a pair $(x, q) \in X \times Q$. This couple consists of continuous $x \in X$ and discrete $q \in Q$ state components.

The properties of hybrid automaton H in Definition 2 mean that, for a μA, its discrete and continuous dynamics can evolve in a free manner. However, jumps in the continuous state space part are not allowed. The previously introduced T-cell model can be derived from this abstract definition taking the output y equal to the state variable x, which is the TCR amount. It is important to note that a *Micro-Agent is a deterministic system*. This means, given an initial condition and an event sequence, that Micro-Agent output is completely defined.

3.5 Stochastic Micro-Agent

In the previous section, the Micro-Agent model is defined. Here, a Stochastic Micro-Agent model will be introduced [53, 56]. First, the Micro-Agent Stochastic Execution is defined. Different assumptions about the MAEG generated sequence lead to different Micro-Agents Stochastic Executions. The concept of stochastic execution [35] is used in the definition of Stochastic Micro-Agents. Particular attention is paid to the case when the MAEG generates events such that the discrete states of a Stochastic Micro-Agent are a Markov Chain.

Definition 3 [35]. (Micro-Agent Stochastic Execution) A stochastic process $(x(t), q(t)) \in X \times Q$ is called a *Micro-Agent Stochastic Execution*, if and only if a Micro-Agent stochastic input event sequence $u(\tau_n)$, $n \in N$, $\tau_0 = 0 \leq \tau_1 \leq \tau_2 \leq \ldots$ generates transitions such that in each interval $[\tau_n, \tau_{n+1})$, $n \in N$, $q(t) \equiv q(\tau_n)$.

Remark 2. The $x(t)$ of a Stochastic Execution is a continuous time function since the transition changes only the discrete state of the Micro-Agent.

Definition 4 [35, 53, 54, 56]. (Micro-Agent Continuous Time Markov Chain Execution) A Micro-Agent Stochastic Execution $(x(t), q(t)) \in X \times Q$ is called a Micro-Agent Continuous Time Markov Chain Execution if the input stochastic event sequence $u(\tau_n)$, $n \in N$, $\tau_0 = 0 \leq \tau_1 \leq \tau_2 \leq \ldots$ generates transitions whose conditional probability satisfies: $P[q(\tau_{k+1}) = q_{k+1} | q(\tau_k) = q_k, q(\tau_{k-1}) = q_{k-1} \ldots q(\tau_0) = q_0] = P[q(\tau_{k+1}) = q_{k+1} | q(\tau_k) = q_k$.

Remark 3. The $q(t)$ of a Micro-Agent Continuous Markov Chain Execution is a Continuous Time Markov Chain.

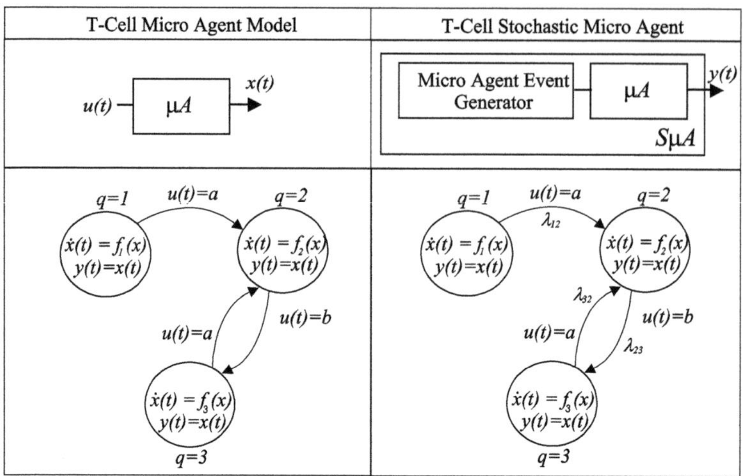

Fig. 3.5. T-cell Micro-Agent and T-cell $CTMC\mu A$ model: continuous state x - the TCR expression level; output $y(t)$ is equal to the state $x(t)$; discrete states q: 1 - *never conjugated*, 2 -*conjugated*, 3 - *free*; events: a - *conjugate formation*, b - *conjugate dissociation*, λ_{ij} - transition rate from the discrete state i to the discrete state j

Definition 5 [53, 54, 56]. (Stochastic Micro-Agent, $S\mu A$) A Stochastic Micro-Agent is a pair $S\mu A = (\mu A, u(t))$, where μA is a Micro-Agent and $u(t)$ is a stochastic input event sequence such that the stochastic process $(x(t), q(t)) \in X \times Q$ is a Micro-Agent Stochastic Execution.

Definition 6 [53, 54, 56]. A Stochastic Micro-Agent ($S\mu A$) is called a Continuous Time Markov Chain Micro-Agent ($CTMC\mu A$) if the input event sequence $u(t)$ is such that the state evolution $(x, q) \in X \times Q$ is a Micro-Agent Continuous Time Markov Chain Execution.

In our definition of a Stochastic Micro-Agent and $CTMC\mu A$ we do not specify the properties of the stochastic event sequence $u(t)$, but in order to apply the definition, this sequence should be characterized.

An example of a $CTMC\mu$ with 3 discrete states, where the sequence $u(t)$ produces the random transitions in such a way that probabilities of the discrete states 1, 2 and 3, P_1, P_2 and P_3, respectively, satisfy the following ODE

$$\begin{bmatrix} \dot{P}_1(t) \\ \dot{P}_2(t) \\ \dot{P}_3(t) \end{bmatrix} = \begin{bmatrix} -\lambda_{12} & 0 & 0 \\ \lambda_{12} & -\lambda_{23} & \lambda_{32} \\ 0 & \lambda_{23} & -\lambda_{32} \end{bmatrix} \begin{bmatrix} P_1(t) \\ P_2(t) \\ P_3(t) \end{bmatrix} \tag{3.3}$$

is illustrated in Figure 3.5. In the left column two equivalent graphical representations of the T-cell Micro-Agent are depicted. The block diagram representation is important in denoting the single-input-single-output nature of the

Micro-Agent, while the state diagram representation is important in describing the internal structure of the Micro-Agent. The right column contains two equivalent descriptions of the $CTMC\mu A$ model. The block diagram denotes that the event sequence generator is the part of the Stochastic Micro-Agent. The state diagram presents the internal structure of the Stochastic Micro-Agent and also the Markov Chain nature of the stochastic transition over the discrete states, denoted by the transition rate diagram.

3.6 Summary

In view of biological facts, we introduce in this section the Micro-Agent model of the T-cell. This is a single-input, single-output hybrid automaton-based model and we use this model as a building block of the T-cell population model. The hybrid automata modeling framework provides us the explicit modeling of cell-to-cell conjugation and dissociation and molecular processes they control. Using the stochastic assumption for the event generation mechanism, the Stochastic Micro-Agent and the $CTMC\mu A$ model of the T-cell in the population are defined. Abstract definitions of the biologically inspired models are introduced using the hybrid automata framework. This section defines the basis for the system-theoretical approach to study Micro-Agent populations which is developed in this work.

Key points

- The proposed model of the population is a collection of Stochastic Micro-Agents.
- The question is how to aggregate their individual behavior in a mathematically tractable way.

4. Micro-Agent Population Dynamics

A mathematical approach to study the relation between the micro- and macro-dynamics of a Micro-Agent population will be developed in this chapter. The approach is motivated by the results of Kinetic Gas Theory [39]. Kinetic Gas Theory deals with a model in which a gas consists of a very large number of small particles in motion. The motivation to exploit this statistical physics reasoning originates from the consideration of a gas as a population of individual particles. A straightforward application of this theory to a general Micro-Agent population is not possible. However, it provides us with an insight on how to make a bridge between the micro-dynamics of the individual Micro-Agent and the Micro-Agent population macro-dynamics.

In Section 4.1, the statistical physics reasoning for the description of the relation between micro- and macro-dynamics of the Micro-Agent population, using a probability density function (PDF), is presented. In Section 4.2, we are concerned with a Continuous Time Markov Chain Micro-Agent ($CTMC\mu A$) stochastic model of the Micro-Agent population and we introduce theorems on the state PDF time evolution. This dynamics is described by a system of partial differential equations (PDE).

4.1 Statistical Physics Background

The Kinetic Gas Theory [39] assumes that a gas is composed of a large number of small particles. One of the most interesting results of this theory, for the problem investigated in this work, is the Maxwell-Boltzmann formula. This formula relates the PDF of the gas particle velocity and the temperature as follows:

$$f(v) = 4\pi \left(\frac{m}{2\pi kT}\right)^{\frac{3}{2}} v^2 exp\left(\frac{-mv^2}{2kT}\right) \tag{4.1}$$

where v is the particle velocity, m is the mass of gas molecules, k is the Boltzmann constant and T is the absolute temperature. The probability density function $f(v)$ relates the dynamics of the individual particle (micro-dynamics), represented by the particle velocity v, to the particle population macro-measurement,

D. Milutinovic and P. Lima: Cells & Robots, STAR 32, pp. 25–34, 2007.
springerlink.com   © Springer-Verlag Berlin Heidelberg 2007

the temperature T. This probability density function(4.1) has two interpretations.

The first interpretation is that equation (4.1) defines the probability density function of the velocity v of one particle. Thus, for a single particle, $f(v)$ defines the probability that this particle will have the velocity v.

Second, if we know the temperature T, equation (4.1) determines the velocity distribution of the particles. This is actually a normalized distribution of the particles over the space of v. Let us suppose that we observe some volume V of the gas with a total number of particles N at time instant t. If we have counted how many particles have the velocity v and this number is $N(v)$, then

$$f(v) = 4\pi \left(\frac{m}{2\pi kT}\right)^{\frac{3}{2}} v^2 exp\left(\frac{-mv^2}{2kT}\right) \simeq \frac{N(v)}{N} \qquad (4.2)$$

The dual meaning of the PDF (4.1) has its roots in the dual nature of the probability. From one perspective, it can be understood as a frequency of some output from repeated experiments. On the other hand, it represents the state of knowledge, or belief, about an experiment output, before an actual observation is made. Regardless of its meaning, the PDF (4.1) contains the complete information about the system, which in a probabilistic way describes the relation between the states of the individual particles and macroscopic observations.

In a very broad sense, the Maxwell-Boltzmann relation inspired our mathematical formulation of the answer to the question:

> *How does the individual Micro-Agent dynamics aggregate to the Macro-Agent population measurements?*

This is the most important step of the theoretical development presented in this monograph. In the formulation of the answer, we are using the PDF of the Micro-Agent state because it is a way to preserve the complete information about the dynamical system state. Using this (complete) information, any kind of the population macro-measurements can be calculated. The proposed answer is:

> *The individual Micro-Agent dynamics and the dynamics of the Micro-Agents population measurements are linked through the probability density function of the Micro-Agent state in that population.*

Although this answer is of a philosophical nature, it provides us with the intuition on how to look for a mathematically tractable answer. In the Kinetic Gas Theory, the PDF (4.1), which connects micro- and macro-dynamics, is inferred using the so-called Maximum Entropy Principle [34]. However, this principle is only valid in the case of the *equilibrium* state of a thermodynamic system. In general case, for $non-equilibrium$ thermodynamics, the Liouville equation [39], which describes the time-evolution of the PDF, must be applied. This equation cannot be applied directly to the state PDF evolution of a Micro-Agent population, where the state is composed of discrete and continuous variables. The result we derive in the following section can be viewed as an extension of the Liouville equation under a Continuous Time Markov Chain assumption about the stochastic Micro-Agent execution (Section 3.5, Definition 4).

4.2 Micro-Agent Population Dynamic Equations

In the previous section, we conclude that the relation between the micro- and macro-dynamics of a Micro-Agent population can be described using the state PDF. Here, this idea will be extended using the assumption that *the complexity of interaction in the population produces a Micro-Agent Stochastic Execution.* This assumption is used due to the complex and unobservable nature of the interaction between individuals. Considering the dual meaning of PDF under this assumption, we can state:

> *The individual Micro-Agent dynamics and the Micro-Agents population measurements dynamics are connected through the probability density function of the Stochastic Micro-Agent state.*

The Stochastic Micro-Agent state PDF represents the state probability of one Micro-Agent. Simultaneously, looking at the population of Micro-Agents, this PDF shows the relative frequency of the state occupancy by individual Micro-Agents. Because of this, we consider a Stochastic Micro-Agent as a stochastic population model.

Relations regarding the Stochastic Micro-Agent state PDF might be difficult to find in general case. However, in the case of $CTMC\mu A$, the system of PDE which describes the time evolution of the state PDF can be found. One of the main outcoming results is presented in the following theorem.

Theorem 1 [53, 54, 56]. For a $CTMC\mu A$ with N discrete states and discrete state probabilities satisfying

$$\dot{P}(t) = L^T P(t) \tag{4.3}$$

where $P(t) = [P_1(t)\ P_2(t)\ldots P_N(t)]^T$, P_i is the probability of discrete state i, $L = [\lambda_{ij}]_{N \times N}^T$, is a transition rate matrix and λ_{ij} is the transition rate from discrete state i to discrete state j, the state PDF is given by the vector

$$\rho(x,t) = [\rho_1(x,t), \rho_2(x,t), \ldots, \rho_N(x,t)]^T \tag{4.4}$$

where $\rho_i(x,t)$ is the PDF of state (x,i) at time t, satisfying the following equation:

$$\frac{\partial \rho(x,t)}{\partial t} = L^T \rho(x,t) - \begin{bmatrix} \nabla \cdot (f_1(x)\rho_1(x,t)) \\ \nabla \cdot (f_2(x)\rho_2(x,t)) \\ \vdots \\ \nabla \cdot (f_N(x)\rho_N(x,t)) \end{bmatrix} \tag{4.5}$$

where

$$\nabla \cdot (f_i(x)\rho_i(x,t)) = \sum_{j=1}^{n} \rho_i(x,t)\frac{\partial f_i^j(x)}{\partial x_j} + f_i^j\frac{\partial \rho_i(x,t)}{\partial x_j} \tag{4.6}$$

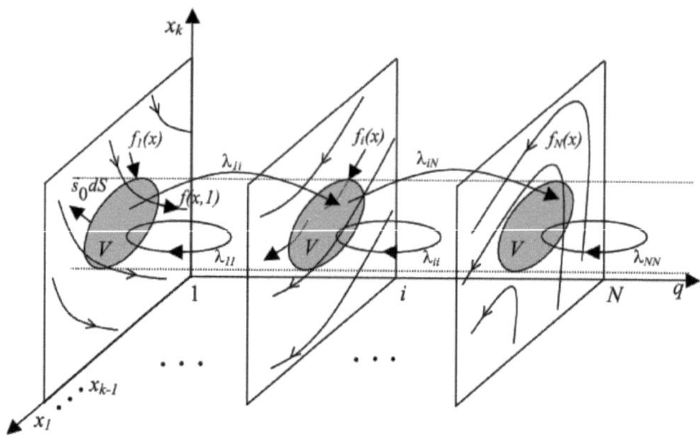

Fig. 4.1. Micro-Agent state space: x_k-kth dimension of the continuous state space, q-discrete state, $f_i(x)$-vector field at $x \in X$ for $q = i$, V-trajectory volume

and $f_i^j(x)$ is the jth component of vector field $f_i(x)$ at state (x, i), $x \in X$, $X = R^n, i = 1, 2, \ldots N$. $\hspace{4cm}$ [*end of theorem*]

Proof. The state space $X \times Q$ of the Stochastic Micro-Agent is illustrated in Figure 4.1. By x_k, we denote the kth dimension of the continuous state space X, q is the discrete state space and $f_i(x)$ is the vector field at $x \in X$ for the discrete state $q = i$.

During the time interval $[\tau_n, \tau_{n+1})$, for the discrete state $q = i$, the Stochastic Micro-Agent dynamics trajectory $x(t)$ evolves according to the differential equation $\dot{x}(t) = f_i(x)$. At the time instant τ_{n+1}, the trajectory jumps to any of the other discrete states with the probability given by (4.3). This jump does not change $x(t)$, since it is a continuous time function, i.e., $x(t^-) = x(t^+) = x(t)$.

The probability $p_{V,i}$ that the Micro-Agent state $(x, q) \in V_i$, $V_i = \{(x, q)|x \in V, q = i\}$ is given by

$$p_{V,i} = \int_V \rho_i(x, t)dV \tag{4.7}$$

where $\rho_i(x, t)$ is the probability density function of the state (x, i) which is inside the arbitrary chosen volume V in X. The time derivative of $p_{V,i}$ is:

$$\dot{p}_{V,i}(t) = \int_V \frac{\partial \rho_i(x, t)}{\partial t} dV \tag{4.8}$$

Figure 4.2 illustrates trajectories in volume V and discrete state i, $i = 1, 2, \ldots, N$.

Let us define the volume V_B, which is a subset of volume V containing the portions of the trajectories crossing the surface S of the volume V in the time interval $[t, t + \Delta t)$. The volume V_B is:

$$V_B = \sum_{S, \Delta S \to 0} f_i(x)\Delta t \cdot \mathbf{s}_0 \Delta S = \oint_S f_i(x)\Delta t dS \tag{4.9}$$

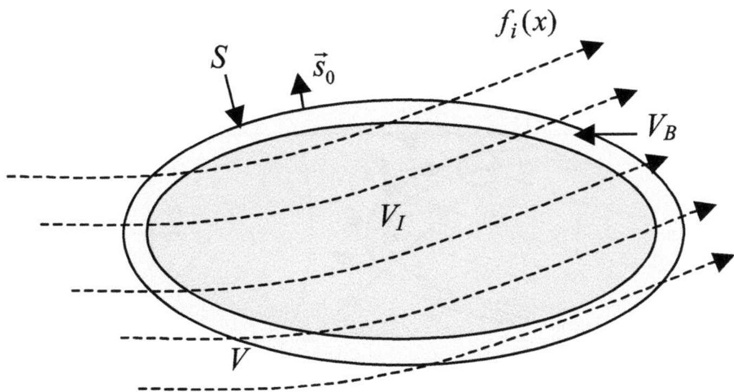

Fig. 4.2. Trajectory volume V: S-surface of the volume V, $\mathbf{s}_0$-vector of the surface S, V_I-volume of the trajectories *not* crossing surface S in the time interval $[t, t + \Delta t)$, V_B-volume of the trajectories crossing surface S in the time interval $[t, t + \Delta t)$

where ΔS is an element of the surface S and $\mathbf{s}_0$ is a vector of this element. The volume V_I in Figure 4.2 is the volume of the trajectory portions which do not leave the volume V, i.e.,

$$V_I = V - V_B \tag{4.10}$$

The probability increase Δp_{V_I} inside the volume V_I during the time interval $[t, t + \Delta t)$, is:

$$\Delta p_{V_I} = \Delta t \sum_{k=1}^{N} \lambda_{ki} p_{V_I,k} = \Delta t \sum_{k=1}^{N} \lambda_{ki} \int_{V_I} \rho_i(x, t) dV \tag{4.11}$$

A portion of the trajectory inside V_I does not leave the volume V, and $x(t)$ is a continuous time function, so there is no change of the probability in V_I due to the vector field $f_i(x)$. However, changes at discrete times can happen due to the Markov Chain transitions and the overall increase Δp_{V_I} is due to these transitions.

To calculate the increase of the probability inside the volume V_B during the time interval $[t, t + \Delta t)$, we will use Figure 4.3. This figure shows an infinitesimally small volume $\Delta V_B = \Delta S \Delta x$. The volume of V_B is given by:

$$V_B = \lim_{\Delta S \to 0} \sum_{S} \Delta S \Delta x \tag{4.12}$$

where ΔS is an infinitely small element of the surface S and Δx, the depth of the volume V_B. The depth of V_B can be calculated to be:

$$\Delta x = \mathbf{s_0} \cdot f_i(x) \Delta t = \Delta t (\mathbf{s_0} \cdot f_i(x)) = \Delta t v \tag{4.13}$$

where v is the $f_i(x)$ projection on the volume V surface vector at element ΔS, $\mathbf{s_0}$. The probability increase inside ΔV_B during the time interval $[t, t + \Delta t)$ at time instant $t + \tau$, $\tau \in [0, \Delta t)$ is:

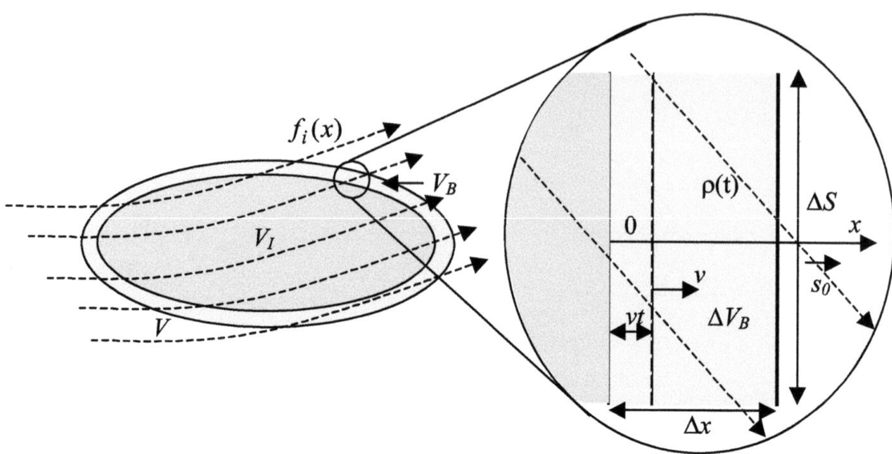

Fig. 4.3. Element of the volume V_B, ΔS - element of the surface S, v - vector field $f_i(x)$ projection on the surface vector $\mathbf{s_0}$, Δx - length $v\Delta t$

$$\Delta p_{\Delta V_B}(t,\tau) = -\rho_i(x,t)\Delta Sv\tau + \dot{\rho}_i(x,t)\tau\Delta S(\Delta x - v\tau) \qquad (4.14)$$

The first term of expression (4.14) presents the probability decrease due to the trajectory outflow. Before the trajectories leave the volume ΔV_B at time $t+\tau$, they produce an increase of the probability due to the random transition over the discrete states. This increase is given by the second term of expression (4.14).

The time derivative of $p_{V,i}$ is:

$$\dot{p}_{V,i}(t) = \lim_{\tau \to 0} \frac{1}{\tau}\left[\Delta p_{V_I} + \sum_{S,\Delta S \to 0} \int_0^\tau \Delta p_{\Delta V_B}(t,s)ds\right] \qquad (4.15)$$

where s is the integration variable. Since the limiting values are:

$$\lim_{\tau \to 0} \frac{1}{\tau}\Delta p_{V_I} = \sum_{k=1}^N \lambda_{ki} \int_{V_I} \rho_k(x,t)dV \qquad (4.16)$$

$$\lim_{\tau \to 0} \frac{1}{\tau}[-\Delta Sv \int_0^\tau \rho_i(x,t)dt] = -\Delta Sv\rho_i(x,t) \qquad (4.17)$$

$$\lim_{\tau \to 0} \frac{1}{\tau} \int_0^\tau \dot{\rho}_i(x,t)\Delta S\Delta x = \lim_{\tau \to 0} \frac{1}{\tau}[\Delta S \sum_{k=1}^N \lambda_{ki} \int_0^\tau \rho_k(x,t)\Delta x dt] = \ldots$$

$$\Delta S\Delta x \sum_{k=1}^N \lambda_{ki}\rho_k(x,t) \qquad (4.18)$$

$$\lim_{\tau \to 0} \frac{1}{\tau} \int_0^\tau \dot{\rho}_i(x,t)(-v\tau) = \lim_{\tau \to 0} \frac{1}{\tau}[\Delta S \sum_{k=1}^N \lambda_{ik} \int_0^\tau \rho_k(x,t)(-vs))ds] = 0 \qquad (4.19)$$

we conclude that

$$\dot{p}_{V,i}(t) = \sum_{k=1}^{N} \lambda_{ki} \int_{V_I} \rho_i(x,t)dV + \sum_{S,\Delta S \to 0} [-\Delta S v \rho_i(x,t) + \Delta S \Delta x \sum_{k=1}^{N} \lambda_{ki}\rho_k(x,t)] \tag{4.20}$$

but, because

$$\sum_{S,\Delta S \to 0} -\Delta S v \rho_i(x,t) = -\oint_S f_i(x)\rho_i(x,t)dS \tag{4.21}$$

and

$$\sum_{S,\Delta S \to 0} \Delta S \Delta x \sum_{k=1}^{N} \lambda_{ki}\rho_k(x,t) = \int_{V_B} \sum_{k=1}^{N} \lambda_{ki}\rho_k(x,t)dV = \sum_{k=1}^{N} \lambda_{ki} \int_{V_B} \rho_k(x,t)dV \tag{4.22}$$

we obtain

$$\dot{p}_{V,i}(t) = \sum_{k=1}^{N} \lambda_{ki} \int_{V_I} \rho_i(x,t)dV + \sum_{k=1}^{N} \lambda_{ki} \int_{V_B} \rho_k(x,t)dV - \oint_S f_i(x)\rho_i(x,t)dS \tag{4.23}$$

i.e.,

$$\dot{p}_{V,i}(t) = \sum_{k=1}^{N} \lambda_{ki} \int_{V} \rho_i(x,t)dV - \oint_S f_i(x)\rho_i(x,t)dS \tag{4.24}$$

Using Gauss' theorem [20]:

$$\dot{p}_{V,i}(t) = \int_{V} \left[\sum_{k=1}^{N} \lambda_{ki}\rho(x,i) - \nabla \cdot (f_i(x)\rho_i(x,t)) \right] dV \tag{4.25}$$

and taking the small volume limit of equations (4.8) and (4.25)

$$\lim_{V \to 0} \dot{p}_{V,i} = \lim_{V \to 0} \int_{V} \frac{\partial \rho_i(x,t)}{\partial t} = \lim_{V \to 0} \int_{V} \left[\sum_{k=1}^{N} \lambda_{ki}\rho(x,i) - \nabla \cdot (f_i(x)\rho_i(x,t)) \right] dV \tag{4.26}$$

we obtain

$$\frac{\partial \rho_i(x,t)}{\partial t} = \sum_{k=1}^{N} \lambda_{ki}\rho_i(x) - \nabla \cdot (f_i(x)\rho_i(x,t)) \tag{4.27}$$

Using $\rho(x,t) = [\rho_1(x),t), \rho_2(x,t), \ldots, \rho_N(x,t)]^T$, the equation system (4.27) becomes

$$\frac{\partial \rho(x,t)}{\partial t} = L^T \rho(x,t) - \begin{bmatrix} \nabla \cdot (f_1(x)\rho_1(x,t)) \\ \nabla \cdot (f_2(x)\rho_2(x,t)) \\ \vdots \\ \nabla \cdot (f_N(x)\rho_N(x,t)) \end{bmatrix} \tag{4.28}$$

Q.E.D

To show that equation (4.2) is an extension of Liouville's equation [39], we will consider the Micro-Agent which has only one discrete state. The state space of such a reduced Micro-Agent is X, instead of $X \times Q$, and by virtue of Theorem 1, the state PDF evolution dynamics reduces to:

$$\frac{\partial \rho(x, y)}{\partial t} = -\nabla \cdot (f(x)\rho(x, t)) \tag{4.29}$$

which is the Liouville equation.

The solution of the partial differential equation (4.2) is the vector of time functions representing the time evolution of the $CTMC\mu A$ state PDF. To solve this equation, the region of interest $\Omega \in X$ and boundary condition should be defined [20]. An example of the boundary condition is $\rho(x, t) = 0$ for all $x \in \partial\Omega$. Specification of the boundary condition depends on the problem which is described by equation (4.2) and can strongly influence the solution [20]. Numerical methods for solving this type of equation are discussed in [85].

The following theorem focuses on is about the time derivatives of the PDE solution:

Theorem 2. For a $CTMC\mu A$ with N discrete states and state probability given by:

$$\dot{P}(t) = L^T P(t) \tag{4.30}$$

where $P(t) = [P_1(t)\ P_2(t) \ldots P_N(t)]^T$, P_i is the probability of the discrete state and $L = [\lambda_{ij}]_{N \times N}^T$, λ_{ij}, a transition rate form discrete state i to discrete state j. The state vector of probability density functions:

$$\rho(x, t) = [\rho_1(x, t)\ \rho_2(x, t) \ldots \rho_N(x, t)]^T \tag{4.31}$$

where $\rho_i(x, t)$, the probability density function of state (x, i) at time t, satisfies:

$$\frac{d\rho(x, t)}{dt} = L^T \rho(x, t) - \begin{bmatrix} \rho_1(x, t)\nabla \cdot f_1(x) \\ \rho_2(x, t)\nabla \cdot f_2(x) \\ \vdots \\ \rho_N(x, t)\nabla \cdot f_N(x) \end{bmatrix} \tag{4.32}$$

where $f_i(x)$ is the vector field value at state (x, i).

Proof. By Theorem 1

$$\frac{\partial \rho_i(x, t)}{\partial t} = \sum_{k=1}^{N} \lambda_{ki}\rho_i(x) - \nabla \cdot (f_i(x)\rho_i(x, t)) \tag{4.33}$$

The total time-derivative of $\rho_i(x, t)$ is:

$$\frac{d\rho_i(x, t)}{dt} = \frac{\partial \rho_i(x, t)}{\partial t} + \nabla\rho_i(x, t) \cdot f_i(x) \tag{4.34}$$

Using equation (4.33) and (4.34), we obtain:

$$\frac{d\rho_i(x,t)}{dt} = \sum_{k=1}^{N} \lambda_{ki}\rho_i(x) - \nabla \cdot (f_i(x)\rho_i(x,t)) + \nabla\rho_i(x,t) \cdot f_i(x) \qquad (4.35)$$

then, by equation (4.6):

$$\nabla \cdot (f_i(x)\rho_i(x,t)) = \nabla\rho_i(x,t) \cdot f_i(x) + \rho_i(x,t)\nabla \cdot f_i(x) \qquad (4.36)$$

Finally,

$$\frac{d\rho_i(x,t)}{dt} = \sum_{k=1}^{N} \lambda_{ki}\rho_i(x) - \rho_i(x,t)\nabla \cdot f_i(x) \qquad (4.37)$$

i.e.,

$$\frac{d\rho(x,t)}{dt} = L^T\rho(x,t) - \begin{bmatrix} \rho_1(x,t)\nabla \cdot f_1(x) \\ \rho_2(x,t)\nabla \cdot f_2(x) \\ \vdots \\ \rho_N(x,t)\nabla \cdot f_N(x) \end{bmatrix} \qquad (4.38)$$

Q.E.D

Theorems 1 and 2 relate to the dynamics of the $CTMC\mu A$ state PDF. Theorem 1 is of special importance. Given an initial state PDF, we can use the theorem to predict the state PDF evolution. In the case of the $CTMC\mu A$ T-cell population model, this theorem can be used to predict the TCR distribution over the T-cell population. It enables us to test different hypotheses about the T-cell micro-dynamics against experimental data, using a mathematical model.

Although the state PDF contains the complete information about population measurements, the experimental test can only be made using the Micro-Agent output PDF $\eta(y,t)$. For an arbitrarily chosen output $y \in Y$, the output PDF is:

$$\eta(y,t) = \int_{V_{x \to y}} \sum_i \rho_i(x,t)dV_{x \to y} \qquad (4.39)$$

where $V_{x \to y}$ is the volume of points $x \in X$ that satisfy $y = g(x)$ (Section 3.4, Definition 2). Unfortunately, solving this integral can be very complex. In the rest of the chapters, we will consider the case when the output vector is equal to the state vector:

$$y = x = [x_1 \ x_2 \ldots x_N]^T \qquad (4.40)$$

in which case, we obtain

$$\eta(y,t) = \eta(y = x, t) = \sum_i \rho_i(x,t), \ i = 1, 2, \ldots N \qquad (4.41)$$

meaning that the output PDF $\eta(y,t)$ can be directly calculated using the state PDFs.

4.3 Summary

The mathematical framework for the study of the relation between the micro-
and macro-dynamics of the Micro-Agent population is developed in this chapter. The approach is based on linking micro- and macro-dynamics using the state
probability density function. Inspired by results of statistical physics, we derive
a PDE system that describes the state probability density function dynamics of
a Stochastic Micro-Agent with a continuous time Markov Chain discrete state
sequence.

Key points

- Through the statistical physics reasoning, the Stochastic Micro-Agent models
  of an individual agent and an agent population are equivalent.
- For the $CTMC\mu A$ model, the evolution of the state probability density function is determined by a system of partial differential equations.

5. Stochastic Micro-Agent Model of the T-Cell Receptor Dynamics

This chapter addresses the TCR expression dynamics investigation using the theoretical results developed in the previous chapters. It is focused on the qualitative and quantitative difference between the results received from the Stochastic Micro-Agent and ODE models of TCR dynamics. Particular attention is paid to the capability of comparing the model-predicted TCR distribution to available experimental data.

The numerical example of Section 5.1 is designed to show how the dynamical equation of the PDF evolution can be applied to the T-cell $CTMC\mu A$ model of the population. In this example, all necessary information about the model is assumed to be based on biological facts, and numerical solutions for three parameter cases are discussed. The difference of the results and the analysis using the $CTMC\mu A$ model instead of the ordinary differential equation (ODE) model, which describes the TCR expression mean value dynamics, is discussed in Section 5.2.

In the first two sections, we discuss technical aspects of using the $CTMC\mu A$ to investigate TCR expression dynamics. However, it is not clear if this model provides any result similar to real data. Using the model structure shown in Figure 5.1 and experimental data, we carry out such a test in Section 5.3. This test shows that the model constraints and experimental data provide us the capability to make inferences about the unknown functions that are part of the individual T-cell TCR dynamics.

In the last section, one part of the $CTMC\mu A$ dynamics is tested against experimental data. This part describes the dynamics of the TCR down-regulation due to the contact with an antibody. We expect this down-regulation to have the same dynamics as the down-regulation during the interaction with the APC. Using the experimental data we identify the parameters of this dynamics.

5.1 T-Cell Receptor Dynamics: A Numerical Example

In this section a numerical example based on the $CTMC\mu A$ population of T-cells (Figure 5.1) is studied. This model is not based on any real data, but

D. Milutinovic and P. Lima: Cells & Robots, STAR 32, pp. 35–52, 2007.
springerlink.com

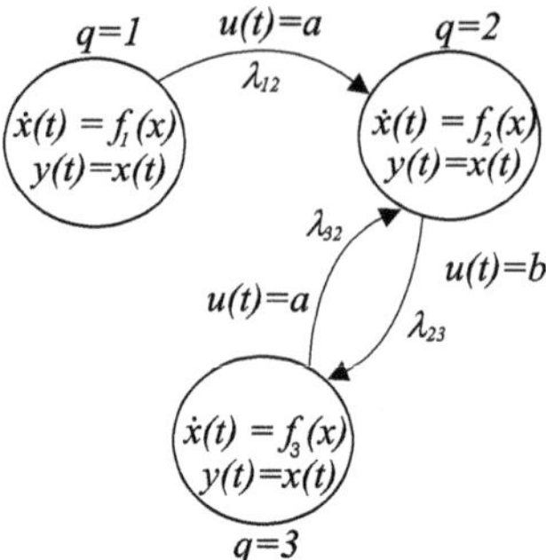

Fig. 5.1. Hybrid automaton model of the T-cell - APC interaction: 1 - *never conjugated*, 2 - *conjugated*, 3 - *free*, a - *conjugate formation*, b - *conjugate dissociation*, λ_{ij} - transition rate from discrete state i to discrete state j [53], ©2003 IEEE

on biological facts and includes the essential properties of the TCR expression dynamics. The numerical example is designed to illustrate the application of Theorem 1, which is derived in Chapter 4.

The Stochastic Micro Agent model of the T-cell population we are considering is already introduced as an illustrative example in Chapter 3. For the purpose of clarity, we present this model again. The $CTMC\mu A$ is a stochastic Micro-Agent model of the T-cell population regarding the TCR expression dynamics. The T-cell can be in one of the three discrete states: *never conjugated, conjugated* and *free*. In each of these states, the TCR expression x changes. Using biological facts, we know that, in the *conjugated* state, the TCR expression decreases and, in the *free* state, the expression increases, so we assume that the dynamics of the increase and decrease can be modeled by $\dot{x} = f_2(x)$ and $\dot{x} = f_3(x)$, respectively. We also assume that the TCR expression of the T-cell, which is *never conjugated* to the APC, remains unchanged, i.e., $\dot{x} = 0$. The output of Micro-Agent y is the TCR expression, i.e., $y = x$. This simplifies the problem of calculating the output PDF $\eta(y, t)$, which is necessary to make a test of the predicted TCR distribution against experimental data. According to this model, each cell in the population changes its discrete state, from state i to state j, with probabilistic rate λ_{ij}.

To make it simpler, all the dynamics in the numerical example presented below are first order linear dynamics:

$$f_1(x, t) = 0; \; f_2(x, t) = -k_2 x; \; f_3(x, t) = k_3 x \tag{5.1}$$

Moreover, to underline that the predictions we make in this section do not relate to experimental data, the TCR expression level is entitled the TCR quantity.

According to Theorem 1, Chapter 4, the probability density function of the state (x, i) satisfies:

$$\frac{\partial}{\partial t}\begin{bmatrix} \rho_1(x,t) \\ \rho_2(x,t) \\ \rho_3(x,t) \end{bmatrix} = \begin{bmatrix} -\lambda_{12} & 0 & 0 \\ \lambda_{12} & -\lambda_{23} & \lambda_{32} \\ 0 & \lambda_{23} & -\lambda_{32} \end{bmatrix}\begin{bmatrix} \rho_1(x,t) \\ \rho_2(x,t) \\ \rho_3(x,t) \end{bmatrix} - \begin{bmatrix} 0 \\ -k_2 x\frac{\partial \rho_2(x,t)}{\partial x} - k_2\rho_2(x,t) \\ k_3 x\frac{\partial \rho_3(x,t)}{\partial x} + k_3\rho_3(x,t) \end{bmatrix}$$

$$(5.2)$$

In sequel, we will numerically solve this system of partial differential equations for the three sets of parameters given in Table 5.1. The parameter values of this numerical example are chosen to illustrate state PDF evolution in the following cases:

- Case I, there is no transitions to *free* state
- Case II, the increase rate k_3 is much lower than the decrease rate k_2
- Case III, the increase rate k_3 is of the same order of magnitude as the decrease rate k_2

In all these cases, the increase rate k_3 is assumed to be lower than the decrease rate k_2.

Table 5.1. T-cell $CTMC\mu A$ model parameter values $[h^{-1}]$

Case	λ_{12}	λ_{23}	λ_{32}	k_2	k_3
I	0.9	0	0.9	0.5	0.25
II	0.9	0.8	0.9	0.5	0.05
III	0.9	0.8	0.9	0.5	0.25

To solve this system of partial differential equations, the domain of the solution, initial and boundary conditions have to be defined [20]. For the domain of the solution, we take the region $0 \leq x \leq 2$. The initial condition is chosen as:

$$\rho_1(x,0) = \begin{cases} 0, & x \leq 0.2 \\ \frac{1}{\sqrt{2\pi\sigma^2}}exp\left(-\frac{(x-1)^2}{2\sigma^2}\right), \sigma = 0.1, & 0.2 < x < 1.8 \\ 0, & x \geq 1.8 \end{cases}$$

$$\rho_2(x,0) = 0 \qquad\qquad (5.3)$$

$$\rho_3(x,0) = 0$$

and the boundary condition is:

$$\begin{bmatrix} \rho_1(0,t) \\ \rho_2(0,t) \\ \rho_3(0,t) \end{bmatrix} = \begin{bmatrix} \rho_1(2,t) \\ \rho_2(2,t) \\ \rho_3(2,t) \end{bmatrix} = \begin{bmatrix} 0 \\ 0 \\ 0 \end{bmatrix} \qquad\qquad (5.4)$$

If the initial and boundary conditions impose contradictory constraints, the solution of the PDE does not exist [20]. To avoid this problem, it is useful to exploit the physical meaning of the PDE problem we are solving [20]. The boundary condition (5.4) imposes that the probability of the T-cell having the TCR quantity $x = 0$ or $x = 2$ is *zero*, for every t which is of interest. Because of that, we shall solve the PDE system for the finite time interval. Considering the parameter values from Table 5.1, any T-cell having the TCR quantity with the *non zero* probability at time $t = 0$, i.e., $x \in [0.2, 1.8]$, cannot reach the $x = 0$ or $x = 2$ in, e.g., $4h$. Therefore, the boundary condition is not in contradiction with the PDE we are solving if we choose $t \in [0, T]$, $T = 4h$.

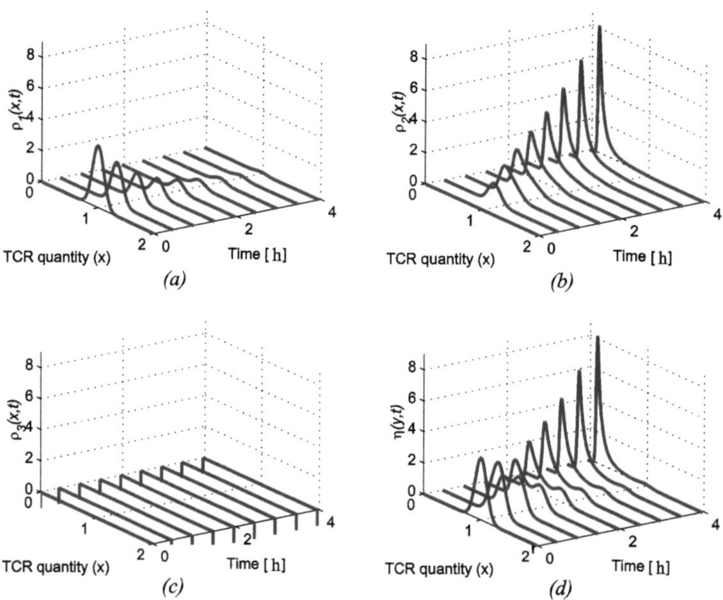

Fig. 5.2. Solution of the PDE system for the T-cell $CTMC\mu A$ model Case I: top-left $\rho_1(x, t)$, top-right $\rho_2(x, t)$, down-left $\rho_3(x, t)$, down-right $\eta(x, t)$; x - TCR quantity, normalized values

Case I (Table 5.1)
The solution of equation (5.2) under conditions (5.3)–(5.4) and parameter Case I (5.1) is shown in Figure 5.2. This figure presents the values of probability density function for each discrete state $\rho_i(x, t)$, $i = 1, 2, 3$, as well as total TCR distribution $\eta(x, t)$ defined as:

$$\eta(x, t) = \rho_1(x, t) + \rho_2(x, t) + \rho_3(x, t) \tag{5.5}$$

This example is interesting because the assumed model has the three discrete states, but with the transition rate from the discrete state $q = 2$ to the discrete state $q = 3$, $\lambda_{23} = 0$. Since the initial condition in the discrete state $q = 3$ is

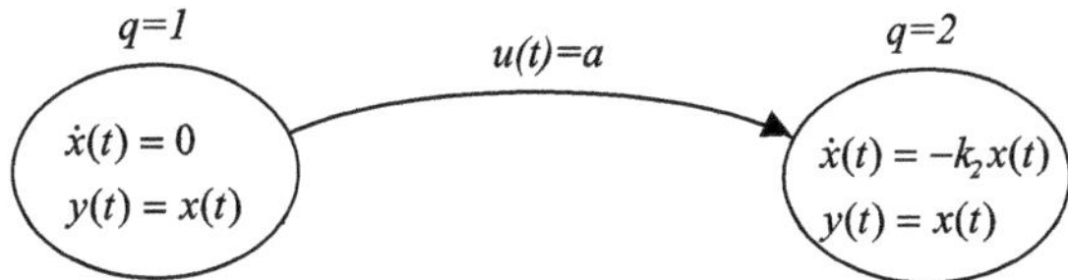

Fig. 5.3. Hybrid Automaton model representing the Micro-Agent model of the T-cell in Case I states: 1 - *never conjugated*, 2 - *conjugated*; event: *a - conjugate formation*

zero, $\rho_3(x,0) = 0$, the probability density function $\rho_3(x,t) = 0, \forall t$. We should notice that the results for $\rho_1(x,t)$ and $\rho_2(x,t)$ are equal to the results calculated for the reduced Micro-Agent presented in Figure 5.3.

Case II (Table 5.1)
This example is different from the previous one, since $\lambda_{23} \neq 0$ and the dynamics in the discrete state $q = 3$ produce an increase of the TCR quantity x. The increase rate k_3 is just 10% of the decrease rate k_2. Despite that, $\rho_2(x,t)$ is flatter than in the Case I. The solution of PDE system (5.2) is presented in Figure 5.4.

Case III (Table 5.1)
In this example, the transition rate between the discrete states $q = 2$ and $q = 3$ is different from zero, $\lambda_{23} \neq 0$, and it is the same as in Case II. However, the increase parameter in the discrete state $q = 3$, k_3 , is 50% of k_2 (parameter for

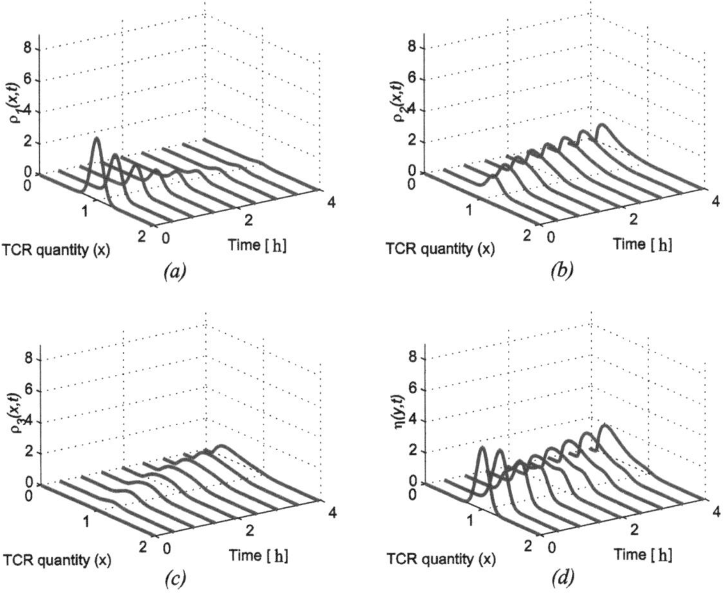

Fig. 5.4. Solution of the PDE system for the T-cell $CTMC\mu A$ model Case II: (a) $\rho_1(x,t)$, (b) $\rho_2(x,t)$, (c) $\rho_3(x,t)$, (d) $\eta(x,t)$; x - TCR quantity, normalized values

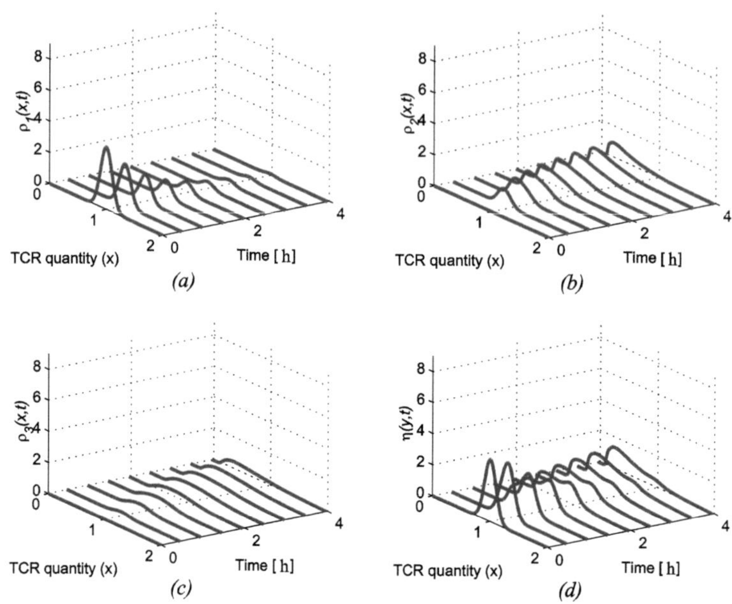

Fig. 5.5. Solution of the PDE system for the T-cell $CTMC\mu A$ model Case III: (a) $\rho_1(x,t)$, (b) $\rho_2(x,t)$, (c) $\rho_3(x,t)$, (d) $\eta(x,t)$; x - TCR quantity, normalized values

the discrete state $q = 2$). This larger parameter k_3 makes $\rho_2(x,t)$ flatter than in Case II. This is because, comparing to Case II, and due to the higher rate of increase k_3, more T-cells in the population have a larger quantity of TCRs. The solution of PDE system (5.2) in Case III is presented in Figure 5.5.

From the results of the numerical example, we can conclude that the solution of the PDE system (5.2) is composed of three functions $\rho_i(x,t), i = 1, 2, 3$, representing the PDF state of T-cell $CTMC\mu A$ model. For the same initial and boundary conditions, the shape of the solution depends on the individual T-cell TCR quantity dynamics parameters and the rate of discrete state transitions. In Section 5.3 and Section 5.4, we will exploit this conclusion to test the hypothesis about individual T-cell TCR expression dynamics based on the TCR expression *steady state* distributions of the T-cell population.

5.2 Micro-Agent vs. Ordinary Differential Equation Model

In this section, we will present the study of a hypothetical T-cell, which is already presented in Section 5.1 as the numerical example (Case I in Table 5.1). We choose this simplest case on purpose to underline the potential differences between results of the ODE models [69] and the $CTMC\mu A$ model of the population TCR expression dynamics.

Let us suppose that biological facts tell us that, after the T-cell APC conjugation, the TCR quantity decreases following the linear ODE law:

$$\dot{x}(t) = -k_2 x(t) \tag{5.6}$$

where x is the quantity of expressed TCRs and k_2 is the reaction rate constant. If this model is valid for all T-cells, then for the known initial TCR average expressed quantity $x_e(0)$, the TCR average quantity x_e can be predicted by the ODE model :

$$\dot{x}_e(t) = -k_2 x_e(t) \tag{5.7}$$

Let us assume now that not only the initial average quantity of the expressed TCRs is known, but also the overall TCR distribution over the T-cell population. If the TCR distribution is normalized, then we get the TCR PDF. To simplify, we assume that the initial TCR PDF is Gaussian with the variance $\sigma^2(0)$. The TCR PDF variance evolution is the solution of the Lyapunov equation [26]:

$$\dot{\sigma}^2(t) = -2k_2\sigma^2(t) \tag{5.8}$$

Since equation (5.6) is linear, the distribution of the TCRs over the T-cell population will be Gaussian at each time instant. The ODE predicted average

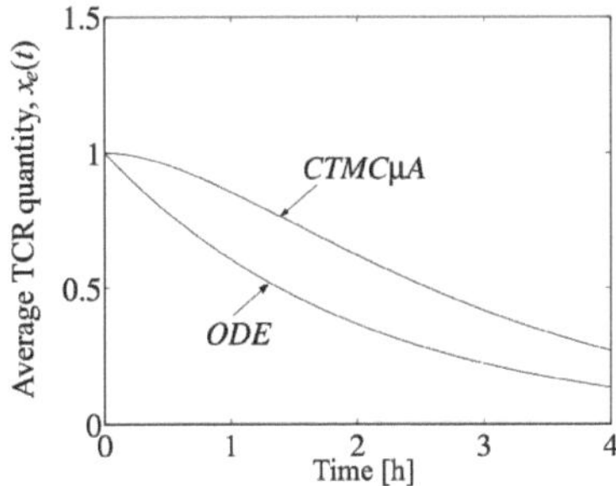

Fig. 5.6. Average value of the TCR quantity over the T-cell population obtained by ODE and Case I $CTMC\mu A$ models (the TCR quantity is normalized)

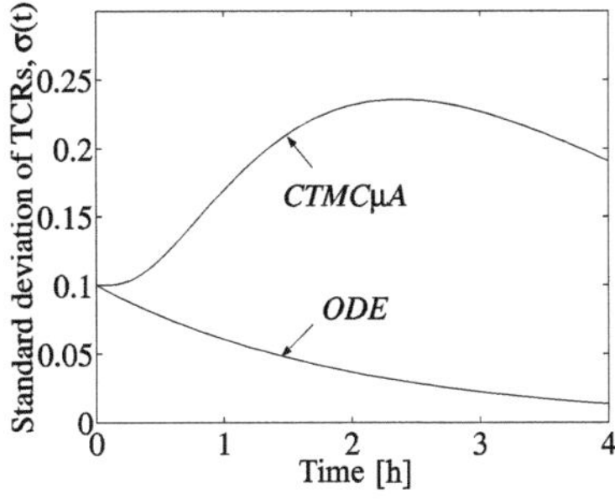

Fig. 5.7. Standard deviation of the TCR quantity over the T-cell population obtained by ODE and Case I $CTMC\mu A$ models (the TCR quantity is normalized)

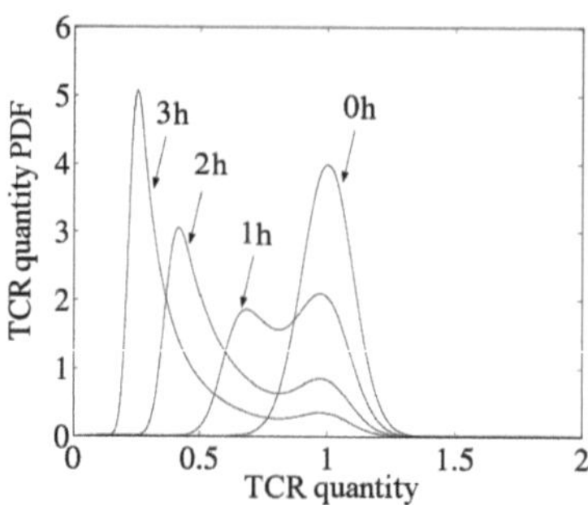

Fig. 5.8. TCR PDF time evolution computed by the $CTMC\mu A$ models (the TCR quantity is normalized)

TCR amount and the standard deviation, for $x_e(0) = 1$, $k_2 = 0.5$ and $\sigma(0) = 0.1$, are presented in Figures 5.6 and 5.7.

The individual T-cell dynamics of TCR decrease in the $CTMC\mu A$ model (Case I, Table 5.1) is the same as in (5.6). However, opposite to the ODE model, in which the TCR decrease starts at the same time instant for all the T-cells, in the $CTMC\mu A$ model we have the rate of the conjugate formation, i.e., the rate of the TCR decrease start. Although the individual T-cell TCR decrease dynamics is the same for both of the models, the $CTMC\mu A$ model predicted average value and the variance do not follow the first order ODE dynamics, due to the rate of the conjugate formation, as shown in Figures 5.6 and 5.7. The effect of the rate of the conjugate formation results in the loss of Gaussian property for the TCR PDF. This is illustrated in Figure 5.8, where the TCR PDF at different time points is presented.

This numerical example illustrates the importance of considering the variance of population measurements in the model identification. Assuming that experimental average values follow the plot corresponding to the $CTMC\mu A$ model in Figure 5.6, this smooth line can be fitted by an ODE model with an appropriate parameter adjustment. However, the associated variance dynamics cannot be explained by the same ODE model. This is because the variance contains the additional information about the individual dynamics of the population. By our Stochastic Micro-Agent modeling approach, we can even go one step further, because we can predict the complete measurement distribution and compare it to the experimentally obtained distribution of measurements.

5.3 T-Cell Receptor Expression Dynamics Model Test

The study of the TCR expression dynamics has motivated the Micro-Agent model development. In the introductory section of this chapter, we propose the T-cell Micro-Agent model (Figure 5.1) based on biological facts. In this section, the theoretically predicted TCR expression distribution will be matched to experimental data [53].

The basis for this analysis is the experimental data in [75]. The experiment therein shows that after long enough time, the TCR distribution remains practically unchanged [75]. This constant distribution will be designated in the sequel by the *steady state* distribution. It is worth mentioning that, due to the lack of data, we are not able to fully justify the application of our model to the data from this experiment, so the steady state assumption might be incorrect. However, our intention here is less ambitious. We would like to see if the model we are considering can produce some reasonable prediction of TCR expression distribution. After we normalize experimentally recorded TCR distribution, so that the integral of the distribution is unity, we can compare it to the T-cell $CTMC\mu A$ output PDF. The role of this example is also to show that, using the steady state TCR expression distribution, we are able to get a further insight into the dynamics of the TCR expression decrease and increase.

Ultimately, the TCR expression level reflects the number, or the amount, of the TCRs expressed on the T-cell surface. To underline that we are dealing with experimental data, we use the term "TCR amount" referring to the TCR expression level.

Let us assume that the transition rates of the proposed T-cell Micro Agent model $\lambda_{12}, \lambda_{23}, \lambda_{32}$ are constant. As explained earlier, $f_1(x,t) = 0$. However, we do not know the continuous dynamics of *conjugated* and *free* states. Therefore, we will assume that they depend only on x and not on t, i.e., the continuous dynamics of the discrete states of the T-cell $CTMC\mu A$ model are

$$f_1(x,t) = 0; \quad f_2(x,t) = f_2(x); \quad f_3(x,t) = f_3(x) \tag{5.9}$$

Due to the reasons explained in Chapter 4, the equation which describes the evolution of the PDF over the $CTMC\mu A$ state space (Theorem 1) is given by [53, 54]:

$$\frac{\partial \rho_1}{\partial t} = -\lambda_{12}\rho_1 - \nabla \cdot (f_1\rho_1) \tag{5.10}$$

$$\frac{\partial \rho_2}{\partial t} = \lambda_{12}\rho_1 - \lambda_{23}\rho_2 + \lambda_{32}\rho_3 - \nabla \cdot (f_2\rho_2) \tag{5.11}$$

$$\frac{\partial \rho_3}{\partial t} = \lambda_{23}\rho_2 - \lambda_{32}\rho_3 - \nabla \cdot (f_3\rho_3) \tag{5.12}$$

where $\rho_i = \rho_i(x,t)$. Since the output $y = x$, the PDF of the $CTMC\mu A$ output $\eta(x,t)$ is given by:

$$\eta(x,t) = \rho_1(x,t) + \rho_2(x,t) + \rho_3(x,t) \tag{5.13}$$

Taking the limit, $t \to \infty$, the steady state of (5.10)-(5.12) is:

$$0 = -\lambda_{12}\rho_1^s - \nabla \cdot (f_1\rho_1^s) \tag{5.14}$$

$$0 = \lambda_{12}\rho_1^s - \lambda_{23}\rho_2^s + \lambda_{32}\rho_3^s - \nabla \cdot (f_2\rho_2^s) \tag{5.15}$$

$$0 = \lambda_{23}\rho_2^s - \lambda_{32}\rho_3^s - \nabla \cdot (f_3\rho_3^s) \tag{5.16}$$

where $\rho_i^s = \rho_i^s(x)$ denotes the steady-state. Since $f_1 = 0$, we can conclude that $\rho_1^s(x) = 0$ and transform the system of equations (5.14)-(5.15) to the equivalent one:

$$0 = -\lambda_{23}\rho_2^s + \lambda_{32}\rho_3^s - \nabla \cdot (f_2 \rho_2^s) \tag{5.17}$$

$$0 = \nabla \cdot (f_2 \rho_2^s + f_3 \rho_3^s) \Leftrightarrow f_2 \rho_2^s + f_3 \rho_3^s = const \tag{5.18}$$

Since the functions ρ_i^s are PDFs, then $\rho_i^s(x) \geq 0$, $\forall x \in R$. The amount of TCRs cannot be negative, thus the PDFs ρ_2^s and ρ_3^s of any negative TCR amount are *zero*. Therefore, there exists a point $x^0 < 0$, where $\rho_2^s(x^0) = \rho_3^s(x^0) = 0$, and equations (5.17)-(5.18) are

$$0 = -\lambda_{23}\rho_2^s + \lambda_{32}\rho_3^s - \nabla \cdot (f_2 \rho_2^s) \tag{5.19}$$

$$0 = f_2 \rho_2^s + f_3 \rho_3^s \tag{5.20}$$

After substituting $\rho_3^s(x) = \eta^s(x) - \rho_2^s(x)$, the solution for the steady state of equations (5.10)-(5.13) is equivalent to the solution of the following ordinary differential equation [53, 54]:

$$\frac{d\eta^s}{dx} = -\left[\left(\frac{f_3 f_2}{f_3 - f_2}\right)^{-1} \frac{d}{dx}\left[\frac{f_3 f_2}{f_3 - f_2}\right] + \frac{\lambda_{23}}{f_2} + \frac{\lambda_{32}}{f_3} \right]\eta^s \tag{5.21}$$

The solution of this equation is [20]:

$$\eta^s(x) = c\left| \frac{1}{f_3(x)} - \frac{1}{f_2(x)} \right| e^{-\int \left(\frac{\lambda_{23}}{f_2(x)} + \frac{\lambda_{32}}{f_3(x)}\right)dx} \tag{5.22}$$

where c has such a value that $\int_\infty^\infty \eta^s(x)dx = 1$. Equation (5.22) defines the shape of the TCR amount distribution. This result is very important because it shows that the TCR amount distribution contains the information about the individual T-cell TCR expression dynamics.

In Figure 5.9, the T-cell - APC interaction experimental data of Valitutti et al. [75] of the initial $\eta(x,0)$ and the steady state $\eta_{exp}^s(x)$ distribution of TCRs over the T-cell population are *approximated by log-normal* distributions [53]:

$$\eta(x,0) = \frac{1}{2\sigma_0 x \sqrt{(\pi)}} e^{-\frac{(M_0 - \ln(x))^2}{2\sigma_0^2}} \tag{5.23}$$

$$\eta_{exp}^s(x) = \frac{1}{2\sigma_\infty x \sqrt{(\pi)}} e^{-\frac{(M_\infty - \ln(x))^2}{2\sigma_\infty^2}} \tag{5.24}$$

where $M_0 = \log_{10}(100)$, $\sigma_0 = 0.19$ and $M_\infty = \log_{10}(50)$, $\sigma_\infty = 0.27$. In order to match steady state distribution (5.22) to equation (5.24), we make the first derivatives in the exponent equal:

$$\frac{\lambda_{23}}{f_2(x)} + \frac{\lambda_{32}}{f_3(x)} = \frac{d}{dx}\frac{(\ln(x) - M_\infty)^2}{2\sigma_\infty^2} \tag{5.25}$$

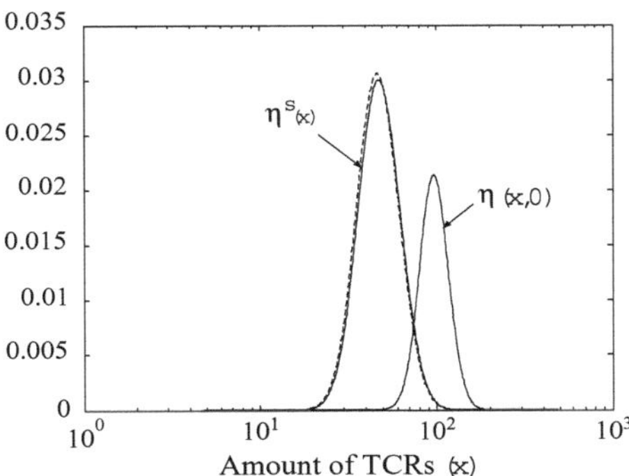

Fig. 5.9. Steady state and initial TCR PDF, based on [75]: $\eta^{s}(x)$ - the $CT\mu A$ model-predicted steady state distribution (*solid*) [53], $\eta_{exp}^{s}(x)$ - normalized smoothed experimental steady state TCR PDF (*dashed*) [53], $\eta(x,0)$ - normalized smoothed experimental initial TCR PDF [53], © 2003 IEEE

After the differentiation in (5.25), we have:

$$\frac{\lambda_{23}}{f_2(x)} + \frac{\lambda_{32}}{f_3(x)} = -\frac{M_\infty}{\sigma_\infty^2}\frac{1}{x} + \frac{1}{\sigma_\infty^2}\frac{\ln(x)}{x} \tag{5.26}$$

The solution of equation (5.26) is composed of $1/x$ and $\ln(x)$ functions, and is not unique. We should take into account that f_2 describes the decrease of TCRs, so we should have $f_2(x) < 0, \forall x > 0$. For the same reason $f_3 > 0, \forall x > 0$, since it describes an increase of TCRs. Following the idea that the increase and decrease dynamics follow different dynamical behavior, one possible solution is [53]:

$$f_2(x) = -k_2 x \tag{5.27}$$

$$f_3(x) = k_3 \frac{x}{\ln(x)} \tag{5.28}$$

where:

$$k_2 = \frac{\sigma_\infty^2 \lambda_{23}}{M_\infty}, k_3 = \sigma_\infty^2 \lambda_{32} \tag{5.29}$$

The previous relations show that the micro dynamics parameters k_2 and k_3 functionally depend on the *conjugate formation* and *conjugate dissociation rates*, λ_{32} and λ_{23}, respectively. If the dissociation rate is higher, then the decreasing rate of TCR, k_2 should be higher, too, i.e., TCR triggering signaling should be more efficient. Similar conclusions can be made about parameter k_3. This functional dependence has been recently reported [13].

Besides the steady state values, the experimental data [75] also contain the time record of the average value of TCRs during the experiment (Figure 5.10). Taking into account the biological fact that the decrease dynamics rate of the

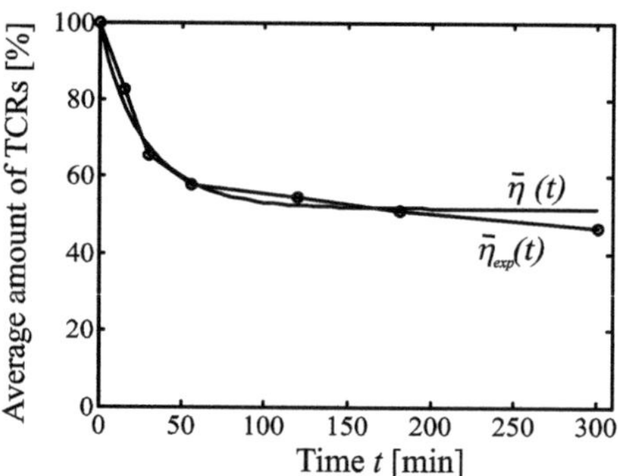

Fig. 5.10. Average amount of TCRs (relative to the initial TCRs amount) [53]: $\overline{\eta}_{exp}(t)$-experimentally obtained values (o) [75], $\overline{\eta}(t)$-the model-predicted average values, © 2003 IEEE

TCRs is much higher than the increase dynamics rate, we assume $k_2 \simeq 100k_3$. To predict higher TCR distribution, the following parameters are chosen:

$$\lambda_{12} = \lambda_{32} = 7, \lambda_{23} = 7000, M_\infty = \log_{10}(50), \sigma_\infty = 0.27 \tag{5.30}$$

The parameters unit is min^{-1}. The predicted steady state TCR distribution $\eta^s(x)$ and the average value $\overline{\eta}^s(t)$ are presented in Figures 5.9 and 5.10, respectively.

We can see that the TCR PDF predicted from the T-cell $CTMC\mu$ model and given parameters cannot be distinguished from the experimental steady state TCR PDF $\eta^s_{exp}(x)$. The predicted average values fit well the experimental average values as well [75].

We do not like to give the impression that the identified model parameters correspond exactly to the parameters of physical reality. We rather conclude that, for proposed mapping of biological knowledge to the T-cell, the Micro-Agent model produces reasonable and meaningful results. This is not obvious from the Micro-Agent model structure and gives us confidence that our line of reasoning of modeling TCR expression dynamics of the T-cell population is reasonable.

5.4 T-Cell Receptor Dynamics in *Conjugated* State

In this section we are analyzing data received from the experiment by Lino [45], which is similar to Valitutti's experiment [75]. In this experiment, the T-cell population is exposed to a large amount of *antibodies*. Antibodies are the soluble molecules which play the role of MHC peptides on the surface of APCs. The T-cell suspension is added to a solution of antibodies and stirred in seconds. Under these conditions, we can assume that all the T-cells start to interact with the antibodies immediately at the beginning of the experiment at time $t = 0$ and

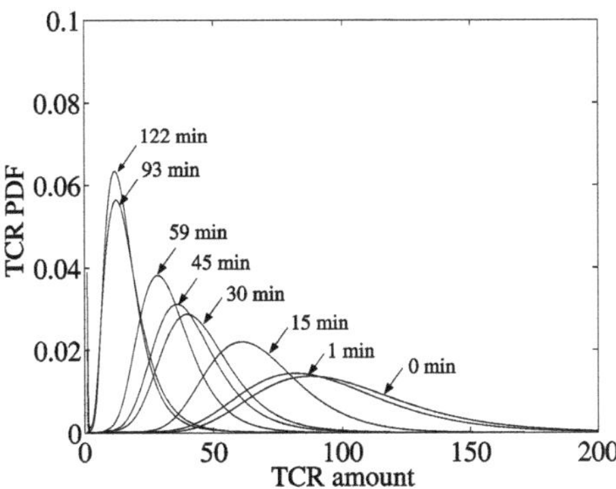

Fig. 5.11. Experimental data TCR PDF estimation [54], $\rho_{exp}(x, t_j)$, $j = 1, 2, \ldots 8$, Flow Cytometry data from [45]

that there is no *conjugate dissociation* afterwards. Thus, all of the T-cells in the population are in the single *conjugated* state. Expecting that the T-cells react in the same way either to APC or antibodies, this experiment provides us a source of data which can be used to identify the dynamics of the individual T-cell in the *conjugated* state [54], i.e., the individual T-cell TCR decrease dynamics.

Since all the T-cells in the population follow dynamics f_2 (see Figure 5.1) of the *conjugated* state, the PDE system, which describes the evolution of the state PDF, reduces to a single PDE, i.e., to the Liouville equation:

$$\frac{\partial \rho(x,t)}{\partial t} = -\nabla \cdot (f_2(x)\rho(x,t)) \tag{5.31}$$

Regarding the individual T-cell TCR dynamics decrease, we pose two hypotheses. The first hypothesis is *linear*:

$$f_2(x) = -k_2 x \tag{5.32}$$

while the second hypothesis is *quadratic*:

$$f_2(x) = -k_2 x^2 \tag{5.33}$$

Both of the hypotheses have been used in the literature, where the ODE dynamics of the average amount of the TCRs is considered [69].

The experimentally estimated TCR PDF $\rho_{exp}(x, t)$ evolution received from our T-cell-antibody experiment is presented in Figure 5.11. The TCR PDF is estimated at $t_1 = 0$, $t_2 = 1$, $t_3 = 15$, $t_4 = 30$, $t_5 = 45$, $t_6 = 59$, $t_7 = 93$ and $t_8 = 122$ min., after the experiment start. Since the TCR PDF contains the important information about the individual T-cell TCR dynamics, as one part of this work, a novel algorithm for estimation of TCR PDF from Flow Cytometry measurements is proposed (see Appendix A). The experimental data presented in Figure 5.11 are produced by this algorithm. The estimated PDF ρ_{exp}, in a log

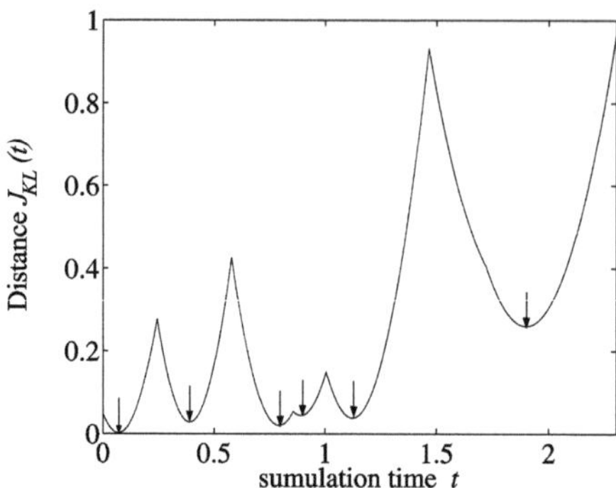

Fig. 5.12. Distance $J_{KL}(t)$ computed for the linear hypothesis [54]. The local minima signaled by the arrows correspond to the time points τ_k when $p(x,t)$ is close to some of the experimental TCR PDF, $p_{exp}(x,t_j)$, $j = 1, 2, \ldots 8$.

scale, is given in Appendix B. The algorithm is based on a stochastic model of the Flow Cytometry measurement of T-cells, QQ-plot PDF estimation [18, 19] and Richardson-Lucy (RL) deconvolution algorithm [47, 66] (see Appendix A).

The first part of the analysis in Section 5.4.1 is concerned with the hypothesis testing using the experimental data plotted in Figure 5.11. The second part of analysis, Section 5.4.2, is concerned with the parameter identification of the individual T-cell TCR expression dynamics.

5.4.1   Model Hypothesis Test

To test the hypotheses of linear and quadratic models of the individual T-cell TCR expression dynamics, we will compare the evolution of the predicted TCR PDF $\rho(x,t)$ using (5.31) the experimentally received TCR PDF $\rho_{exp}(x,t_j)$, $j = 1, 2, \ldots 8$ [54]. The initial condition for the prediction is $\rho(x,0) = \rho_{exp}(x,0)$. The shape of the evolution $\rho(x,t)$, calculated by (5.31) at different times, does not depend on parameter k_2, either for the linear or for the quadratic hypotheses. It is because the parameter k_2 in equation (5.31), for both *linear* and *quadratic* case, only scales the time. This allows us to take $k_2 = 1$ and compare the shapes in the prediction $\rho(x,t)$ to the shapes in the experimental data $\rho_{exp}(x,t_j)$. Using $k_2 = 1$, we should keep in mind that t and t_j are not measured within the same time frame.

However, to compare the shape of the prediction $\rho(x,t)$ to the experimental data $\rho_{exp}(x,t_j)$, we need to choose the times τ_k, $k = 1, 2, \ldots$ at which we are comparing $\rho(x,\tau_k)$ to $\rho_{exp}(x,t_j)$. We determine τ_k as the times when the model predicted PDF $\rho(x,t)$ is the most similar to some of the experimentally received PDF $\rho_{exp}(x,t_j)$, $j = 1, 2, \ldots 8$. To find τ_k, we use the distance which compares the PDFs in terms of the Kullback-Leibler [KL] distance [15]:

$$J_{KL}(\rho(t), \rho_{exp}) = \min_j \sum_i \rho(x_i, t) \log \frac{\rho(x_i, t)}{\rho_{exp}(x_i, t_j)} \tag{5.34}$$

measuring the distance between the predicted PDF at time t, $\rho(x,t)$ and the set of experimentally received measurements ρ_{exp} [54], computed as a minimal value of KL distance between $\rho(x,t)$ and $\rho_{exp}(x,t_j)$, $j = 1, 2, \ldots 8$. The Kullback-Leibler distance is the entropy-based distance commonly used to measure the similarity between the two PDFs. The distance $J_{KL}(\rho(t), \rho_{exp})$ computed for the linear hypothesis PDF prediction is presented in Figure 5.12. It is small whenever $\rho(x,t)$ is similar to some of the $\rho_{exp}(x,t_j)$, $j = 1, 2, \ldots 8$. Therefore, the time points τ_k, $k = 1, 2, \ldots$ correspond to the local minima of $J_{KL}(\rho(t), \rho_{exp})$, which are signaled by the arrows in Figure 5.12.

The TCR PDF evolution for the linear hypothesis, calculated by (5.31) at the time points τ_k, $k = 1, 2, \ldots 6$, is plotted in Figure 5.13. From the figure, we can see that the linear hypothesis shape of $\rho(x,t)$ evolution corresponds to the shape of the experimental TCR PDF pretty well.

Similar analysis can be made for the quadratic hypothesis. However, this makes little sense, since in the quadratic case, the TCR PDF evolution does not

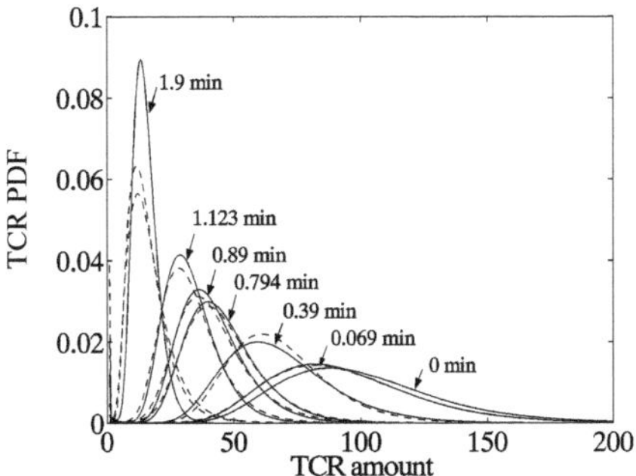

Fig. 5.13. Time evolution of TCR PDF, linear hypothesis [54]: model-predicted $\rho(x,t)$(*solid*, computed at time τ_k), the experimental TCR PDF $\rho_{exp}(x,t_j)$ (*dashed*), $j = 1, 2, \ldots 8$, Flow Cytometry data from [45]

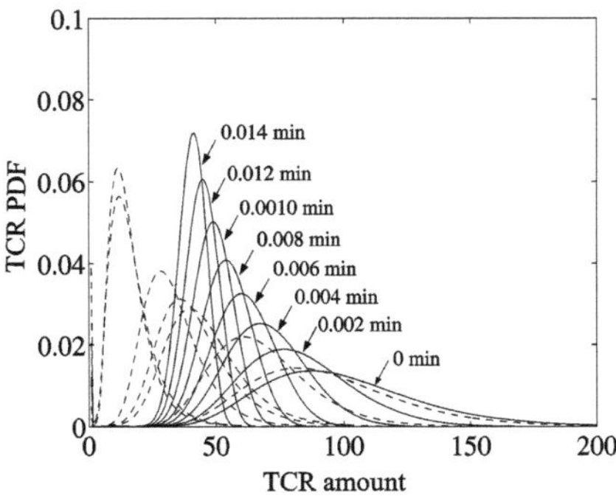

Fig. 5.14. Time evolution of TCR PDF, quadratic hypothesis [54]: model-predicted $\rho(x,t)$(*solid*), the experimental TCR PDF $\rho_{exp}(x,t_j)$ (*dashed*), $j = 1, 2, \ldots 8$, Flow Cytometry data from [45]

produce the shapes of $\rho(x,t)$ which approximate closely the experimental TCR PDF, Figure 5.14.

Based only on the shapes in $\rho(x,t)$ evolution, we can conclude that the individual T-cell TCR expression dynamics of the *conjugated state* is closer to the linear hypothesis. This conclusion corresponds to the result of the analysis presented in the previous section.

5.4.2 Parameter Identification

In Figures 5.13 and 5.14, the TCR PDF prediction and the experimental TCR PDF are not compared under the same time frame. This is allowed because our hypothesis test takes into account only the shape of the model predicted TCR PDF evolution, which does not depend on the parameter k_2. In the following analysis, we will estimate this parameter comparing $\rho(x,t)$ and $\rho_{exp}(x,t_j)$, where t and t_j are measured under the same time frame.

We first introduce a cost function J which depends on k_2:

$$J(k_2) = \sum_{t_j} \sum_{i} \rho(x_i, t_j | k_2) \log \frac{\rho(x_i, t_j | k_2)}{\rho_{exp}(x_i, t_j)} \tag{5.35}$$

where $\rho(x_i, t | k_2)$ is the value of $\rho(x,t)$ calculated for the parameter k_2 at $x = x_i$, and $\rho_{exp}(x_i, t_j)$ is the value of $\rho_{exp}(x, t_j)$ at $x = x_i$. This function is a sum of KL-based distances between the predicted PDF at time t_j, for the given parameter k_2 and the experimentally observed $\rho_{exp}(x_i, t_j)$.

The minimum of the function $J(k_2)$ means that the predicted PDF evolution of $\rho(x, t | k_2)$ is the closest possible to the experimentally received PDFs ρ_{exp} at each sampling instant t_j. The function $J(k_2)$ computed for the linear hypothesis and experimental data is presented in Figure 5.15. Using this figure, we can conclude that $k_2 = 0.015$. The corresponding time evolution of $\rho(x, t | k_2)$ is presented in Figure 5.16. The reason why the TCR PDF prediction, $k_2 = 0.015$,

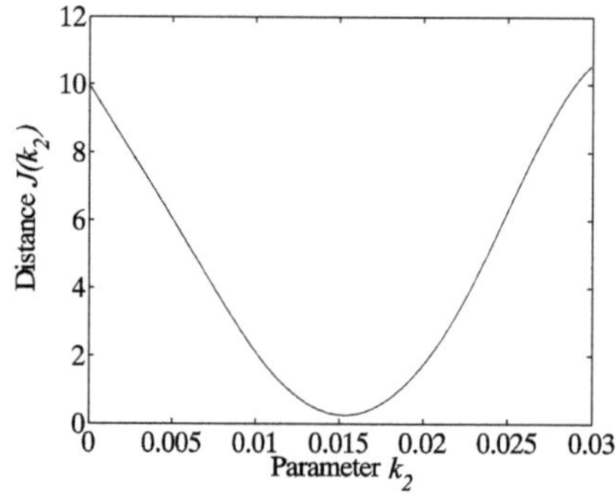

Fig. 5.15. Distance between the model prediction and experimental data against the parameter k_2

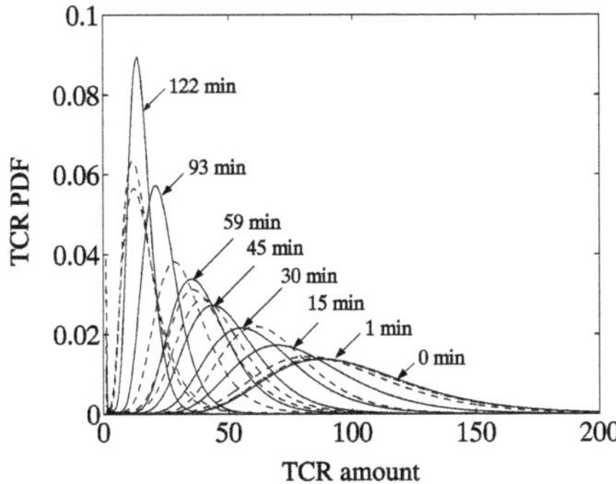

Fig. 5.16. Predicted TCR PDF for $k_2 = 0.015$(*solid line*), experimental TCR PDF $\rho_{exp}(x, t_j)$ (*dashed*), $j = 1, 2, \ldots 8$, Flow Cytometry data from [45]

does not match better the experimental TCR PDF can be ascribed to the following reasons:

- The individual T-cell TCR dynamics is more complex than the time invariant linear one. For example, the parameter k_2 depends on time.
- The experimental TCR PDFs are estimated incorrectly due to the incorrect stochastic measurement model.
- The experimental TCR PDFs are estimated using very similar, but not identical, T-cell populations. It can be the reason why the experimentally received distributions are interlaced with the predicted TCR PDFs. Because of the same reason, the experimental distributions at $93rd$ min and $122nd$ min (see Figure 5.11 and Figure 5.16 (dashed)) are equally close to the prediction at $122nd$ min (Figure 5.16, solid).

Investigation of these sources of uncertainty and understanding of their influence on the difference between predicted and experimental data is essential for future work.

At the end of this section, it is worth saying that in the case of the linear decrease dynamics and when only one discrete state exists, the evolution of the PDF mean value and variance can be described by the corresponding ODE model. Thus, the parameter k_2 can be obtained from fitting the data to the ODE model. However, in the case of nonlinear hypothesis, the complete machinery used to measure the distance between the time evolution of predicted and experimentally received PDFs has to be applied. Our analysis shows that making the prediction of TCR expression distribution measurements and comparing these distributions to experimental data can be very efficient in eliminating any wrong hypothesis about TCR expression dynamics.

5.5 Summary

The application of a $CTMC\mu A$ stochastic micro agent model to the investigation of individual T-cell TCR expression dynamics is presented in this chapter.

We propose the $CTMC\mu A$ T-cell model and we show that the TCR PDF predicted by this model contains the information about the individual T-cell TCR dynamics. Making hypothesis about the individual dynamics, we find that our $CTMC\mu A$ can explain experimental data well and we conclude that the TCR dynamics decrease of an individual T-cell, due to the contact of the T-cell to the APC or antibodies, is almost linear.

Using a $CTMC\mu A$ instead of an ODE model, we can match the TCR distribution to the experimental data instead of matching only the average TCR amount. Matching distributions places strong constraints on the hypothesis for the TCR expression dynamics of the individual T-cell. It provides a qualitatively better insight into the individual T-cell TCR expression dynamics, which is based on the experimentally received TCR distribution of the T-cell population.

Key points

- The numerical example illustrates an application of mathematical development of $CTMC\mu A$ mathematical developments.
- The $CTMC\mu A$ model is employed to explain the steady state in T-cell-APC experimental data of Valitutti et al. [75].
- The dynamics of *conjugated state* is investigated comparing the evolution of the predicted TCR PDF to the experimental data from the T-cell-antibody experiment by Lino [45].

6. Stochastic Micro-Agent Model Uncertainties

The T-cell population models presented in Chapter 5 are based on the simplified assumption that the continuous dynamics of all individual Micro-Agents of the population is identical. Motivated by the problem of modeling the diversity in biological interactions, we describe here a method of dealing with the parameter uncertainty of the $CTMC\mu A$ continuous dynamics. This method is focused on a systematic procedure that modifies the original $CTMC\mu A$ state PDF dynamics equations to a set of equations that incorporate the parameter uncertainty. Our discussion in this chapter is limited to the case of a $CTMC\mu A$ single parameter uncertainty. However, the suggested procedure is general enough to be applied to more complex cases, involving several uncertain parameters.

In the case of the T-cell $CTMC\mu A$ model (Chapter 4), this single parameter is the rate of the TCR expression decrease while the T-cell and the APC are conjugated. This parameter is uncertain because it depends on the APC characteristics. Different APCs expose different amounts of MHC-peptide complexes on their surfaces. Although the parameter is uncertain, it is constant while the cells are conjugated. This kind of the $CTMC\mu A$ parameter uncertainty will be analyzed below.

From the $CTMC\mu A$ definition (Chapter 3), the continuous dynamics f_i are defined for each discrete state $i \in Q$, where Q is the set of $CTMC\mu A$ discrete states. In the single parameter uncertainty case, among all discrete states Q, there exists only one discrete state (say, $w \in Q$) with the continuous dynamics (f_w) dependent on an uncertain scalar parameter b. This parameter b is unknown, but constant, during the time intervals when the $CTMC\mu A$ is in the discrete state w. Each time the $CTMC\mu A$ enters the state w, the parameter b can have a new constant random value. Assuming that the parameter b is a scalar-valued random variable, defined by its PDF $p(b)$, two types of parameter uncertainty can be distinguished:

- The *discrete parameter uncertainty*, for which the parameter b is random, but the values it can take are from a finite set $\{b_1, b_2, \ldots b_D\}$, where D is the number of all the possible parameter values, see Figure 6.1a.
- The *continuous parameter uncertainty*, for which the parameter b is random and it takes the value from some range in R^1; and the PDF $p(b)$ is a continuous function, see Figure 6.1b.

D. Milutinovic and P. Lima: Cells & Robots, STAR 32, pp. 53–66, 2007.
springerlink.com

The PDF examples in the case of the *discrete parameter uncertainty* (DP uncertainty) and the *continuous parameter uncertainty* (CP uncertainty) are presented in Figure 6.1. We will analyze both uncertainty cases in the following sections.

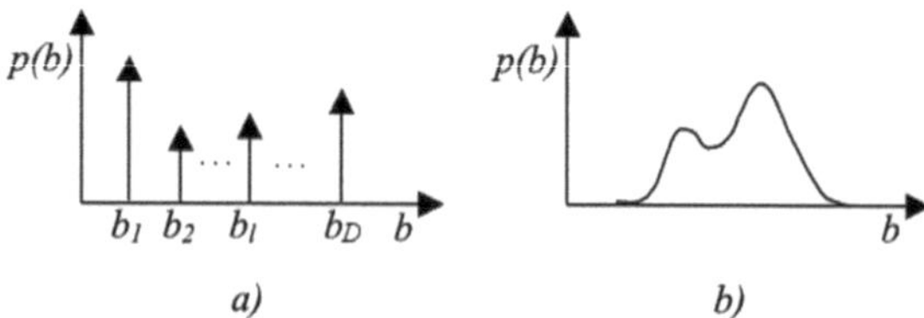

Fig. 6.1. Two cases of the parameter uncertainty: (*a*) The PDF of the discrete parameter uncertainty, b_d-parameter values, $d = 1, 2, \ldots D$, D-the number of possible values. (*b*) The PDF of the continuous parameter b.

6.1 Discrete Parameter Uncertainty Case

To explain the methodology that we develop, the definitions of input and output state sets of a discrete state are introduced first. Then, the basic idea of our method is explained and we derive a theorem which states the conditions for the invariance of the discrete state probability under the discrete state splitting of a given Continuous Markov Chain. Using this theorem, the method that modifies the original $CTMC\mu A$ and generates a new $CTMC\mu A$ model, that incorporates the parameter uncertainty (denoted by $\overline{CTMC\mu A}$), is introduced.

Definition 6. Given a Continuous Markov Chain, with the transition matrix $L^T = [\lambda_{ij}]_{N \times N}$, where $N = |Q|$, Q is the set of discrete states and λ_{ij}, the transition rate from state $i \in Q$ to state $j \in Q$, we define:

- $In(w)$ - the *set of input states of the discrete state w* by

$$In(w) = \{q | \lambda_{qw} > 0, q \in Q\} \tag{6.1}$$

- $Out(w)$ - the *set of output states of the discrete state w* by

$$Out(w) = \{q | \lambda_{wq} > 0, q \in Q\} \tag{6.2}$$

In other words, the set $In(w)$ is the set of discrete states from which the state w can be reached after one transition. Similarly, $Out(w)$ denotes the set of the discrete states that can be reached from the discrete state w after one transition. In Figure 6.2a, this part of a $CTMC\mu A$ is illustrated. The discrete state w is the one which contains the uncertain parameter b, while the states $j \in In(w)$ and $k \in Out(w)$ are chosen from non-empty $In(w)$ and $Out(w)$ sets.

The basic idea of modifying the $CTMC\mu A$ to the $\overline{CTMC\mu A}$ is depicted in figure 6.2b. Let us assume that the $CTMC\mu A$ has just the vector field f_w, which depends on $b \in \{b_1, b_2, \ldots, b_D\}$. This means, that after transition to the state w,

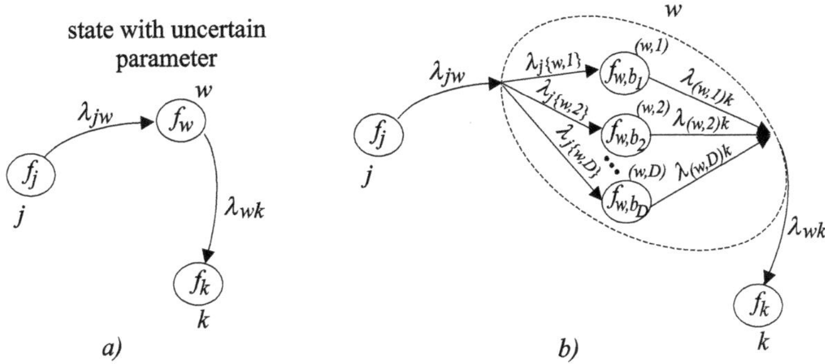

Fig. 6.2. $CTMC\mu A$ modification: (a) part of the original $CTMC\mu A$ (b) the splitting of the state w, which is the state with the parameter uncertainty

the trajectory of the $CTMC\mu A$ changes in one of D directions, given by f_{w,b_d}, $d \in \{1, 2, \ldots D\}$. The direction to follow is chosen based on the parameter b, which is a discrete-valued random variable determined by its PDF $p(b)$. Since $p(b)$ is a set of Dirac impulses, the *probability* of the b_d realization is denoted by $p(b_d)$ to simplify the notation (reserving the capital letter P for discrete state probabilities).

If we split the discrete state w into the states (w, d), each of them assigned to one vector field f_{w,b_d}, this results in a $\overline{CTMC\mu A}$, which incorporates the b parameter uncertainty of the original $CTMC\mu A$ model. However, to make this modification correctly, the input rate $\lambda_{j(w,d)}$ and the output transition rate $\lambda_{(w,d)k}$ of the introduced states have to be determined so that the modification does not influence the rest of discrete state probabilities. In other words, for the rest of the discrete state dynamics, the state w division has to be "invisible". The discrete dynamics of a $CTMC\mu A$ is described by a Markov Chain, which satisfies:

$$\dot{P}(t) = L^T P(t) \tag{6.3}$$

where $P(t)$ is the vector $(P(t) = [P_1, \ldots P_w \ldots P_N]^T)$ of discrete state probabilities and L is the transition rate matrix. Mathematically, the state w splitting is "invisible" if the complete evolution of $P(t)$ is invariant under this modification. The following theorem provides the conditions for this invariance to hold for a continuous time Markov Chain.

Theorem 1. If the evolution of the Continuous Markov Chain discrete state probabilities vector $P(t)$ is given by:

$$\dot{P}(t) = L^T P(t) \tag{6.4}$$

where $L = [\lambda_{ij}]$ is the matrix of transition rates from the discrete state $i \in Q$ to the discrete state $j \in Q$ and one of the discrete states, w, is split into (w, d),

$d = 1, 2, \ldots D$ discrete states, then under this modification, $P(t)$ is invariant $\forall \, P(0) \in R^n$ and $\forall t$ if all the following conditions hold:

(A)

$$P_w(0) = \sum_{d=1}^{D} P_{(w,d)}(0) \tag{6.5}$$

(B)

$$\lambda_{jw} = \sum_{d=1}^{D} \lambda_{j(w,d)} \tag{6.6}$$

(C)

$$\sum_{k=1}^{|Out(w)|} \lambda_{(w,d)k} = |Out(w)|\lambda_{wk} \tag{6.7}$$

where $|Q|$ is the cardinality of the discrete state set Q.

Proof. After the discrete state w is decoupled, $P(t)$ is invariant if

$$P_w(t) = \sum_{d=1}^{D} P_{(w,d)}(t) \tag{6.8}$$

(A) The equation (6.8) should be satisfied at $t = 0$, because of:

$$P_w(0) = \sum_{d=1}^{D} P_{(w,d)}(0) \tag{6.9}$$

(B) The probability increase of P_w, which results from the transition from state $j \in Q$ and is denoted by $\Delta P_{w|j}(t)$, is given by:

$$\Delta P_{w|j}(t) = \lim_{\Delta t \to 0} \lambda_{jw} P_j(t) \Delta t \tag{6.10}$$

because $P_{w|j}(t)$ is the conditional probability increase of P_w due to the state j. Using the same reasoning, the increase of the probability of the discrete state (w, d), due to the transition from the state $j \in Q$, $P_{(w,d)|j}$, is given by:

$$\Delta P_{(w,d)|j}(t) = \lim_{\Delta t \to 0} \lambda_{j(w,d)} P_j(t) \Delta t \tag{6.11}$$

For $P(t)$ to be invariant, $P_w(t)$ must be invariant and also its increase must be invariant, i.e.,

$$\Delta P_{w|j}(t) = \sum_{d=1}^{D} \Delta P_{(w,d)|j}(t) \tag{6.12}$$

which requires:

$$\lambda_{jw} = \sum_{d=1}^{D} \lambda_{j(w,d)} \tag{6.13}$$

that proves the condition (B) of the theorem, equation (6.6).

(C) Let us write the expression for the probability increase of the discrete state w. According to the theorem statement, we have:

$$\dot{P}_w(t) = \sum_{j \in In(w)} \lambda_{jw} P_j(t) - \sum_{k \in Out(w)} \lambda_{wk} P_w(t) \tag{6.14}$$

We write the set of equations which describe the time increase of the state PDF corresponding to the introduced D discrete states $(w, \boldsymbol{d})$:

$$\dot{P}_{(w,1)}(t) = \sum_{j \in In(w)} \lambda_{j(w,1)} P_j(t) - \sum_{k \in Out(w)} \lambda_{(w,1)k} P_{(w,1)}(t) \tag{6.15}$$

$$\dot{P}_{(w,2)}(t) = \sum_{j \in In(w)} \lambda_{j(w,2)} P_j(t) - \sum_{k \in Out(w)} \lambda_{(w,2)k} P_{(w,2)}(t) \tag{6.16}$$

$$\cdots \quad \cdots \tag{6.17}$$

$$\dot{P}_{(w,D)}(t) = \sum_{j \in In(w)} \lambda_{j(w,D)} P_j(t) - \sum_{k \in Out(w)} \lambda_{(w,D)k} P_{(w,D)}(t) \tag{6.18}$$

Having satisfied (6.8), the following also holds:

$$\dot{P}_w(t) = \sum_{d=1}^{D} \dot{P}_{(w,d)}(t) \tag{6.19}$$

and, using equations (6.15)-(6.18), this time derivative is:

$$\dot{P}_w(t) = \sum_{d=1}^{D} \left[\sum_{j \in In(w)} \lambda_{j(w,d)} P_j(t) - \sum_{k \in Out(w)} \lambda_{(w,d)k} P_{(w,d)}(t) \right] \tag{6.20}$$

Making this equation equal to (6.14), we have

$$\sum_{d=1}^{D} \left[\sum_{k \in Out(w)} \lambda_{(w,d)k} P_{(w,d)}(t) - \sum_{j \in In(w)} \lambda_{j(w,d)} P_j(t) \right] + \dots$$

$$\sum_{j \in In(w)} \lambda_{jw} P_j(t) - \sum_{k \in Out(w)} \lambda_{wk} P_w(t) = 0 \tag{6.21}$$

and by re-arranging summation in this expression, we obtain:

$$\sum_{j \in In(w)} \lambda_{jw} P_j(t) - \sum_{k \in Out(w)} \lambda_{wk} P_w(t) - \sum_{j \in In(w)} P_j(t) \sum_{d=1}^{D} \lambda_{j(w,d)} + \dots$$

$$\sum_{d=1}^{D} \left[\sum_{k \in Out(w)} \lambda_{(w,d)k} P_{(w,d)}(t) \right] = 0 \tag{6.22}$$

From the proven condition (B), the following equation is true:

$$- \sum_{k \in Out(w)} \lambda_{wk} P_w(t) + \sum_{d=1}^{D} \left[\sum_{k \in Out(w)} \lambda_{(w,d)k} P_{(w,d)}(t) \right] = 0 \tag{6.23}$$

in view of equation (6.8):

$$\sum_{k \in Out(w)} \lambda_{wk} \left[\sum_{d=1}^{D} P_{(w,d)}(t) \right] - \sum_{d=1}^{D} \left[\sum_{k \in Out(w)} \lambda_{(w,d)k} P_{(w,d)}(t) \right] = 0 \qquad (6.24)$$

Collecting together all the results dependent on k, we finally obtain:

$$\sum_{d=1}^{D} \left[\sum_{k \in Out(w)} (\lambda_{(w,d)k} - \lambda_{wk}) \right] P_{(w,d)}(t) = 0 \qquad (6.25)$$

This equation has to be satisfied for all t, including $t = 0$ and all the initial conditions $P_{(w,i)}(0)$. This is possible only if

$$\sum_{k \in Out(w)} \lambda_{(w,d)k} = \sum_{k \in Out(w)} \lambda_{wk} = |Out(w)| \lambda_{wk} \qquad (6.26)$$

which proves the condition (C) of the theorem, (6.7). $\qquad$ Q.E.D.

We should notice that Theorem 6.1 conditions are actually the constraints on how the set of input $(\lambda_{j(w,d)})$ and output $(\lambda_{j(w,d)})$ transition rates and initial conditions $(P_{(w,d)})$ ($d = 1, 2, \ldots D$) have to be chosen. The available degrees of freedom in the parameter selection can be used for appropriate modeling purposes.

The same degree of freedom does not exist if the input transition rates are not dependent on d. In that case we have:

$$\lambda_{j(w,d)} = \lambda_{jw} p(b_d), \quad j \in In(w) \qquad (6.27)$$

which satisfies Theorem 6.1. Similarly, for the output transition independent of d, the only thing we can conclude is that:

$$\lambda_{(w,1)k} = \lambda_{(w,2)k} = \ldots = \lambda_{(w,2)D}, \quad k \in Out(w) \qquad (6.28)$$

Thus, if the $P(t)$ is invariant under the state w splitting, the application of the Theorem 6.1 condition (C) yields:

$$\lambda_{(w,d)k} = \lambda_{wk}, \quad d = 1, 2, \ldots D \qquad (6.29)$$

Theorem 6.1 is valid for Markov Chains and it can be, therefore, applied to the Markov Chain which is embedded in the $CTMC\mu A$. However, in the case of the original $CTMC\mu A$, we have:

$$P_w(0) = \int_X \rho_w(x, 0) dx \qquad (6.30)$$

where $\rho_w(x, 0)$ is the initial PDF of the $CTMC\mu A$ in the discrete state w. After the state w decomposition, we have:

$$P_w(0) = \sum_{d=1}^{D} P_{(w,d)}(0) = \sum_{d=1}^{D} \int_X \rho_{(w,d)}(x, 0) \qquad (6.31)$$

Table 6.1. Constraints for the $CTMC\mu A$ to $\overline{CTMC\mu A}$ modification in the case of the *discrete parameter uncertainty*: w - the $CTMC\mu A$ discrete state with parameter b dependence ($b \in \{b_1, b_2, \ldots b_D\}$), $\lambda_{jw}, \lambda_{wk}$ and $\rho_w(x,0)$ - the input transition rate, the output transition rate and the initial PDF at discrete state w; (w, d) - the introduced discrete states of the $\overline{CTMC\mu A}$, $\lambda_{j(w,d)}$, $\lambda_{(w,d)k}$ and $\rho_{(w,d)}(x,0)$ - the input transition rate, the output transition rate and the initial PDF at discrete state (w, d); $j \in In(w)$, $k \in Out(w)$, $d = 1, 2, \ldots D$

	Parameter dependent	Parameter independent
Input transition rates	$\lambda_{jw} = \sum_{d=1}^{D} \lambda_{j(w,d)}$	$\lambda_{j(w,d)} = \lambda_{jw} p(b_d)$
Output transition rates	$|Out(w)|\lambda_{wk} = \sum_{k \in Out(w)} \lambda_{(w,d)k}$	$\lambda_{wk} = \lambda_{(w,d)k}$
Initial probability	$\rho_w(x,0) = \sum_{d=1}^{D} \rho_{(w,d)}(x,0)$	$\rho_{(w,d)}(x,0) = \rho_w(x,0) p(b_d)$

i.e., the condition (A) of Theorem 6.1 is satisfied only if:

$$\rho_w(x,0) = \sum_{d=1}^{D} \rho_{(w,d)}(x,0)dx \tag{6.32}$$

If $\rho_{w,d}(x,0)$ does not depend on parameter b, then:

$$\rho_w(x,0) = \sum_{d=1}^{D} \rho_{(w,d)}(x,0), \quad d = 1, 2 \ldots, D \tag{6.33}$$

which is satisfied because $\sum_{d=1}^{D} p(b_d) = 1$

Table 6.1 presents the conditions that must be satisfied to split the discrete state w into the new discrete states (w, d). In this table, the cases, when the input and output transition rates and the initial probability are *dependent* and when they are *independent* on the uncertain parameter b, are listed in the separate columns. The previous discussion about handling the discrete parameter uncertainty in the state w is summarized in the three-step procedure listed bellow.

The three-step procedure for including the Discrete Parameter uncertainty into the $CTMC\mu A$ model

Step 1) Apply Theorem 1 (Chapter 4) to the $CTMC\mu A$ as if it is the case without uncertainty.

Remark: The ith row of the PDE system that corresponds to the discrete state $i \in Q$ is:

$$\frac{\partial \rho_i(x,t)}{\partial t} = \sum_{j=1}^{N} \lambda_{ji} \rho_j(x,t) - \nabla \cdot (f_i(x)\rho_i(x,t)) \quad i = 1, 2, \ldots N \tag{6.34}$$

Step 2) Split the discrete state w, which depends on the parameter b, into the discrete states (w, b_d), which correspond to one of the possible realizations of the parameter $b \in \{b_1, b_2, \ldots D\}$, $d = 1, 2, \ldots D$. Then apply the results in Table 6.1 and update the transition rates. This substitution produces the $\overline{CTMC\mu A}$.

Remark: The corresponding PDE system is

$$\frac{\partial \rho_i(x,t)}{\partial t} = \sum_{j=1,j\neq w}^{N} \lambda_{ji}\rho_j(x,t) - \sum_{d=1}^{D} \lambda_{(w,d)i}\rho_{(w,d)}(x,t) - \nabla \cdot (f_i(x)\rho_i(x,t)) \quad (6.35)$$

for all $i \neq w$ and instead of one equation for w, we now have D equations for the discrete states $(w, \boldsymbol{d})$

$$\frac{\partial \rho_{(w,\boldsymbol{d})}(x,t)}{\partial t} = \sum_{j=1,j\neq w}^{N} \lambda_{j(w,\boldsymbol{d})}\rho_j(x,t) + \sum_{k=1,j\neq w}^{N} \lambda_{(w,\boldsymbol{d})k}\rho_{(w,\boldsymbol{d})}(x,t) - \ldots$$
$$\nabla \cdot (f_{w,b_{\boldsymbol{d}}}(x)\rho_{(w,\boldsymbol{d})}(x,t) \quad (6.36)$$

Step 3) Re-index the discrete state, so that the new index includes the following discrete states:

$$1, 2, \ldots, w - 1, (w, 1), (w, 2), \ldots, (w, D), w + 1, \ldots, N \quad (6.37)$$

The total number of the $\overline{CTMC\mu A}$ discrete states is now $N + D - 1$.

The above procedure explains how to deal with the discrete parameter uncertainty which appears in a single discrete state. If the parameter b is a vector, then the original model must be extended with the number of discrete states which correspond to the number of all possible vector values given their PDF $(p(b))$. In the case when this type of uncertainty appears in more than one state, each discrete state with uncertainty must be substituted by the discrete states corresponding to each possible parameter value. The input and output rates have to be arranged in the same way as it is explained above. The easiest way to handle this case is to apply the *three step procedure* only to a single state with uncertainty, obtain the associated $\overline{CTMC\mu A}$, and then, with this model, apply the same procedure on the next discrete state with uncertainty. Thus, we conclude that the derived *three-step procedure* is general enough to be applied in all cases of the $CTMC\mu A$ with discrete parameter uncertainties.

6.2 Continuous Parameter Uncertainty Case

The approach to continuous parameter uncertainty of a $CTMC\mu A$ is completely based on the result derived in the previous section. We assume that the uncertain parameter b takes the value from a range $B \in R^1$ with a probability given by the PDF $p(b)$. The development is based on the discretization of the range B. It produces an $CTMC\mu A$ with an infinite number of discrete states. In the limit, as the discretization step approaches zero, a system of equations, which incorporates the continuous parameter b, is derived. Using this system of equations, the

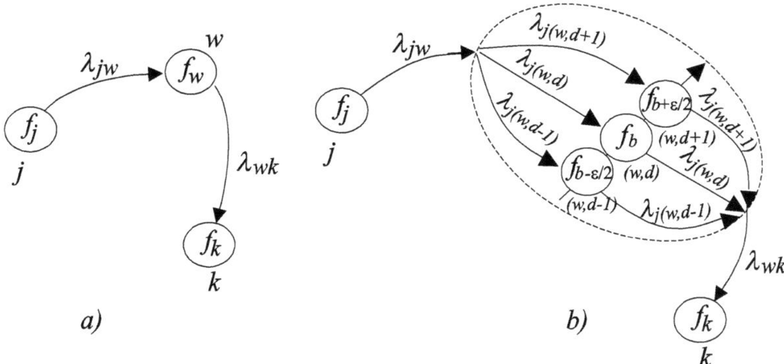

Fig. 6.3. $CTMC\mu A$ modification: (a) part of the original $CTMC\mu A$ (b) the state w division using the parameter b discretization, w - the state with the parameter uncertainty, $\epsilon - discretization\ step$

systematic procedure to transform the $CTMC\mu A$ state PDF dynamics equation to a set of equations, which incorporates the parameter uncertainty, is obtained.

Figure 6.3a presents the original $CTMC\mu A$. The discrete state w is the one in which the continuous dynamics depends on the parameter b. By discretizing the range B, we can split the state w into the discrete states $(w, \boldsymbol{d})$, $\boldsymbol{d} = 1, 2, \ldots D$, each of them assigned to the parameter $b \in (b_{\boldsymbol{d}} - \epsilon/2, b_{\boldsymbol{d}} + \epsilon/2]$, where ϵ is the discretization step. The system of equations that describes the $CTMC\mu A$ state PDF is the same as the one in *step 2* of the *three-step procedure* from the previous section. Denoting explicitly the continuous state space of the operator ∇ by ∇_x, the *step 2* equations are :

$$\frac{\partial \rho_i(x,t)}{\partial t} = \sum_{j=1, j \neq w}^{N} \lambda_{ji} \rho_j(x,t) - \sum_{d=1}^{D} \lambda_{(w,d)i} P_{(w,d)}(x,t) - \ldots$$
$$\nabla_x \cdot (f_i(x)\rho_i(x,t)), \ i \neq w \qquad (6.38)$$

$$\frac{\partial \rho_{(w,d)}(x,t)}{\partial t} = \sum_{j=1, j \neq w}^{N} \lambda_{j(w,d)} \rho_j(x,t) + \sum_{k=1, k \neq w}^{N} \lambda_{(w,d)k} P_{(w,d)}(x,t) - \ldots$$
$$\nabla_x \cdot (f_{w,b_{\boldsymbol{d}}}(x)\rho_{(w,d)}(x,t)) \quad (6.39)$$

Taking into account that:

$$\lambda_{j(w,d)} = \int_{b_{\boldsymbol{d}}-\epsilon/2}^{b_{\boldsymbol{d}}+\epsilon/2} \lambda_{jw}(b)db \ \text{ and } \ \lambda_{(w,d)i} = \int_{b_{\boldsymbol{d}}-\epsilon/2}^{b_{\boldsymbol{d}}+\epsilon/2} \lambda_{wi}(b)db \qquad (6.40)$$

equations (6.38) and (6.39) can be written as:

$$\frac{\partial \rho_i(x,t)}{\partial t} = \sum_{j=1, j \neq w}^{N} \lambda_{ji} \rho_j(x,t) - \sum_{d=1}^{D} \int_{b_{\boldsymbol{d}}-\epsilon/2}^{b_{\boldsymbol{d}}+\epsilon/2} \lambda_{wi}(b) \rho_{(w,d)}(x,t)db - \ldots$$
$$\nabla_x \cdot (f_i(x)\rho_i(x,t)), \ i \neq w \quad (6.41)$$

$$\frac{\partial \rho_w(x, b_d, t)}{\partial t} = \sum_{j=1, j \neq w}^{N} \lambda_{jw}(b_d) \rho_j(x, t) + \ldots$$

$$\sum_{k=1}^{N} \lambda_{wi}(b_d) \rho_{(w,d)}(x, t) db - \nabla_x \cdot (f_{w, b_d}(x) \rho_w(x, b_d, t)) \tag{6.42}$$

In equation (6.42), the following substitution is made:

$$\rho_{(w,d)}(x, t) = \rho_w(x, b_d, t) \tag{6.43}$$

which is a simple change of the notation, where the index d is substituted with the corresponding parameter value b_d, which is the parameter of ρ_w. Taking into account that:

$$\lim_{\epsilon \to 0} \sum_{d=1}^{D} \int_{b_d - \epsilon/2}^{b_d + \epsilon/2} \lambda_{wi}(b) db = \int_B \lambda_{wi}(b) db \tag{6.44}$$

and taking the limit $\epsilon \to 0$ of equations (6.41) and (6.42) that also corresponds to $D \to \infty$ and $b_d \to b$, we obtain :

$$\frac{\partial \rho_i(x, t)}{\partial t} = \sum_{j=1, j \neq w}^{N} \lambda_{ji} \rho_j(x, t) - \int_B \lambda_{wi}(b) \rho_w(x, b, t) db - \ldots$$

$$-\nabla_x \cdot (f_i(x) \rho_i(x, t)), \ i \neq w \tag{6.45}$$

$$\frac{\partial \rho_w(x, b, t)}{\partial t} = \sum_{j=1, j \neq w}^{N} \lambda_{jw}(b) \rho_j(x, t) + \sum_{k=1, k \neq w}^{N} \lambda_{wi}(b) \rho_w(x, b, t) - \ldots$$

$$\nabla_x \cdot (f_{w,b}(x) \rho_w(x, b, t)), \ \forall b \in B \tag{6.46}$$

This system of PDE equations describes the $CTMC\mu A$ PDF dynamics in the presence of the parameter with the continuous parameter uncertainty of b. *This is a system with an* infinite number of equations *because the integral in the equation (6.45) can be solved only if the equation (6.46) is solved for each value of the parameter $b \in B$*. One way to approach this problem is to approximate the integral in (6.46) and to solve a large finite number of the equations (6.46). The transition rates and the initial condition in the previous equations have to satisfy the conditions given in Table 6.2. This table includes the cases when the input and/or the output transitions of the state w are dependent or independent on the parameter b.

The way to deal with the continuous parameter uncertainty presented above represents the case when the uncertainty appears in a single discrete state. If this type of uncertainty appears in more than one state, then the set of equations (6.45) has as many integral terms as the discrete states with uncertainty. To each integral term, i.e., to each discrete state with uncertainty, a system of equations of the type (6.46), is assigned.

Table 6.2. Constrains on the equation set that incorporates the continuous parameter uncertainty: w - the $CTMC\mu A$ discrete state with the parameter $b \in B$ $\lambda_{jw}, \lambda_{wk}$ and $\rho_w(x, 0)$ - the input transition rate, the output transition rate and the initial PDF at the discrete state w; $\lambda_{jw}(b)$, $\lambda_{wk}(b)$ and $\rho_w(x, b, 0)$ - parameters and the initial condition of the new set of equations that incorporates uncertainty. $j \in In(w)$, $k \in Out(w)$

	Parameter dependent	Parameter independent
Input transition rates	$\lambda_{jw} = \int_B \lambda_{jw}(b)db$	$\lambda_{jw}(b) = \lambda_{jw}p(b)$
Output transition rates	$\|Out(w)\|\lambda_{wk} = \sum_{k \in Out(w)} \lambda_{wk}(b)$	$\lambda_{wk} = \lambda_{wk}(b)$
Initial probability	$\rho_w(x, 0) = \int_B \rho_w(x, b, 0)db$	$\rho_w(x, b, 0) = \rho_w(x, 0)p(b)$

6.3 Numerical Example

In this section, we will apply the theory developed in this chapter to the TCR triggering dynamics model of Section 5.1. To clarify, we depict this model again in Figure 6.4. To this model, we apply the assumption that the TCR dynamics of a T-cell, while it is conjugated to the APC, is identical for all the T-cells in the population. However, the intensity of the individual T-cell TCR internalization in the T-cell-APC conjugate depends on how many MHC-peptide complexes are present on the surface of the APC. Because of that, we should take into account the parameter uncertainty of the TCR dynamics in the *conjugated* discrete state.

The model depicted in Figure 6.4 is also used in Section 5.3 for the analysis of experimental data. There we concluded that TCR PDF is equivalent to the *normalized* TCR distribution, and that steady state TCR PDF and dynamics of the average TCR amount can be explained using the following continuous dynamics of discrete states (Section 5.3, Equations (5.27)-(5.29)) :

$$f_2(x) = -k_2 x , \quad f_3(x) = k_3 \frac{x}{\ln(x)} \tag{6.47}$$

where:

$$k_2 = \frac{\sigma_\infty^2 \lambda_{23}}{M_\infty} , \quad k_3 = \sigma_\infty^2 \lambda_{32} \tag{6.48}$$

and

$$\lambda_{12} = \lambda_{32} = 7, \lambda_{23} = 7000, M_\infty = \log_{10}(50), \sigma_\infty = 0.27 \tag{6.49}$$

The parameter unit is min^{-1}. The predicted steady state TCRs PDF $\eta^s(x)$ and average value $\overline{\eta}(t)$, using these parameter values and equations from Appendix C, are presented in Figures 6.6 and 6.7, respectively.

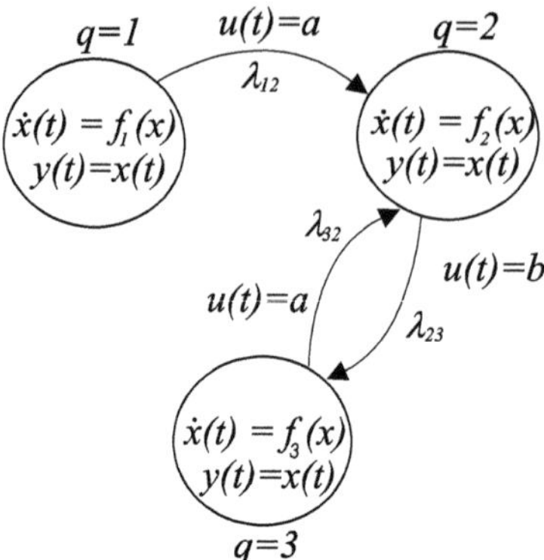

Fig. 6.4. *$CTMC\mu A$ model of the T-cell - APC interaction: 1 -never conjugated, 2 -conjugated, 3 -free, a - conjugate formation, b - conjugate dissociation* [53], © 2003 IEEE

The uncertain parameter of *conjugated* discrete state is k_2. To describe its uncertainty, the values and the probability of k_2 realization, when the T-cell enters the *conjugated* discrete state, must be specified. We will assume that uncertain parameter k_2 can take the three discrete values k_2^1, k_2^2, k_2^3 with the probability:

$$p(k_2^1) = p(k_2^2) = p(k_2^3) = \frac{1}{3} \tag{6.50}$$

This reflects our assumption that all the three values are equally probable when T-cell enters the *conjugated* state.

Applying the results summarized in Table 6.1, we can derive the model $\overline{CTMC\mu A}$, which includes the parameter k_2 uncertainty, presented in Figure 6.5. This model has five discrete states. The states $(2,1)$, $(2,2)$ and $(2,3)$ are the result of the state 2 splitting, and the dynamics of each state resulting from state 2 are $f_2^1(x)$, $f_2^2(x)$ and $f_2^3(x)$, respectively. Based on Table 6.1, we can find that the transition rates of the $\overline{CTMC\mu A}$ model are:

$$\lambda_{1(2,1)} = \lambda_{1(2,2)} = \lambda_{1(2,1)} = \lambda_{12}/3 \tag{6.51}$$

$$\lambda_{(2,1)3} = \lambda_{(2,2)3} = \lambda_{(2,3)3} = \lambda_{23} \tag{6.52}$$

$$\lambda_{3(2,1)} = \lambda_{3(2,2)} = \lambda_{3(2,1)} = \lambda_{32}/3 \tag{6.53}$$

Using the equations for the $\overline{CTMC\mu A}$ model (Figure 6.5), given in Appendix C, we can predict the steady state TCR PDF and the average value of the TCRs for the two cases of parameter k_2 uncertainty, where $k_2^2 = k_2$. In the first case, parameter values k_2^1 and k_2^3 deviate 10% from the value of k_2, which means that $k_2^1 = 0.9k_2$ and $k_2^3 = 1.1k_2$. In the second case, parameter values deviate 20% from the value of k_2, meaning that $k_2^1 = 0.8k_2$ and $k_2^3 = 1.2k_2$. The results of the prediction for these two cases will have superscript 10% and 20%, respectively.

The prediction of the steady state TCR PDF $\eta^{10\%}(x)$ and $\eta^{20\%}(x)$ are presented in Figure 6.6 and the average amount TCR predictions $\bar{\eta}^{10\%}(t)$ and

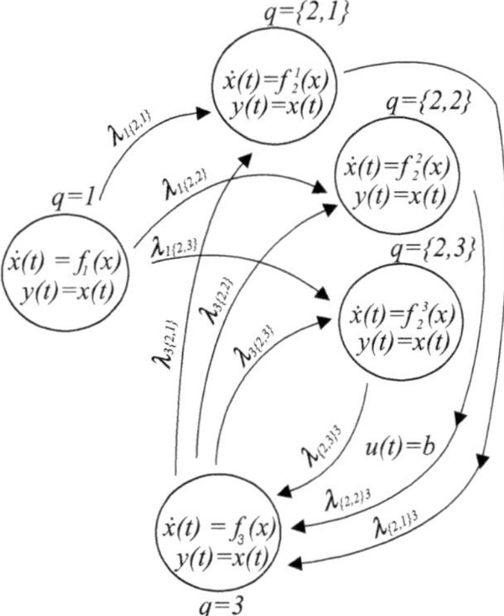

Fig. 6.5. $\overline{CTMC\mu A}$ model of the T-cell - APC interaction, which includes the parameter k_2 uncertainty: 1 - *never conjugated*; $(2, i)$ - *conjugated* with the dynamics $f_2^i(x)$ corresponding to parameter value $k_2^i, i = 1, 2, 3$; 3 -*free*; *a - conjugate formation*; *b - conjugate dissociation*

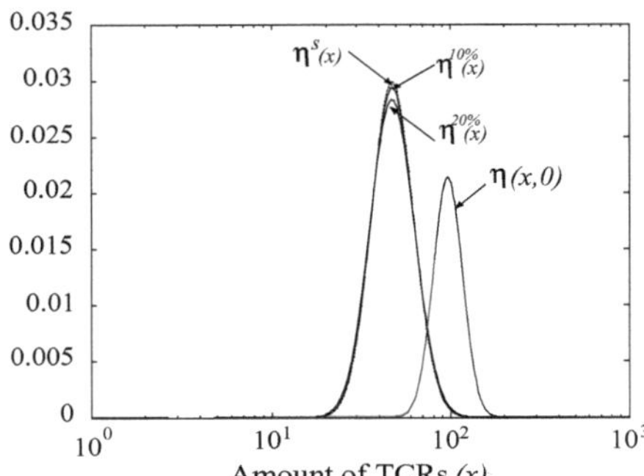

Fig. 6.6. Steady state and the initial TCR PDF, based on [75]: $\eta^s(x)$ - the $CTMC\mu A$ model predicted steady state distribution (*dotted*) [53]; $\eta(x,0)$ - normalized smoothed initial TCR PDF [53]; $\eta^{10\%}(x)$ - steady state prediction with 10% deviation of parameter values k_2^1 and k_2^3; $\eta^{20\%}(x)$ - steady state prediction with 20% deviation of parameter values k_2^1 and k_2^3, © 2003 IEEE

$\bar{\eta}^{20\%}(t)$ can be seen in Figure 6.7. From these results, we can see that the prediction based on the original model without uncertainty, $CTMC\mu A$, and on the model with uncertainty, $\overline{CTMC\mu A}$, are slightly different. Therefore, we can conclude that predictions based on the original model are not very sensitive to the k_2 parameter uncertainty in this case.

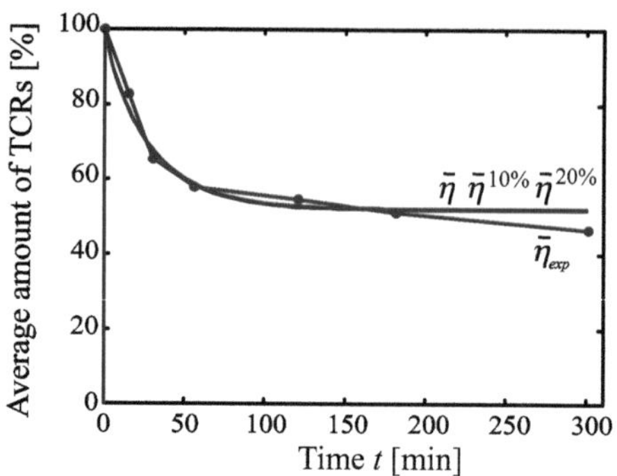

Fig. 6.7. Average amount of TCRs (relative to the initial TCR amount) [53]: $\overline{\eta}_{exp}(t)$-experimentally obtained values (o) [75], $\overline{\eta}(t)$-the model predicted average values, $\overline{\eta}^{10\%}(t), \overline{\eta}^{20\%}(t)$-the model predicted average values with 10% and 20% deviation of parameter values k_2^1 and k_2^3, © 2003 IEEE

6.4 Summary

In this section we present an approach to dealing with parameter uncertainties of the $CTMC\mu A$ continuous dynamics with the parameter uncertainty described by its PDF. Both discrete and continuous parameter uncertainties are considered. The method to handle the parameter uncertainty is developed by the modification of the original $CTMC\mu A$ model to the $\overline{CTMC\mu A}$ model, which incorporates the parameter uncertainty. In the case of a discrete parameter, the *three step procedure*, which yields this modification, is derived. In the continuous parameter uncertainty case, the same approach leads us to a system of partial differential equations with the integral terms that include state variables.

Key point

- The *three step procedure* that modifies the original $CTMC\mu A$ model to the $\overline{CTMC\mu A}$ model which includes *discrete parameter uncertainty*.

7. Stochastic Modeling and Control of a Large-Size Robotic Population

By studying the problem of modeling biological systems, a more general approach to study a system composed of a large population of individuals is developed. We find this approach general enough to provide results of potential interest in engineering applications concerning Multi-Agent systems (MAS). The main motivation of our approach is to provide fundamental principles of modeling and control for large-size populations, providing math-based tools for the analysis and control design from specifications.

Here we assume that the robotic population can execute primitive tasks driven by individual robot controllers and the local information. For example, each primitive task may correspond to one motion primitive. We introduce a model of robotic population which is based on the Micro-Agent modeling approach, leading to a system of partial differential equations (PDE) that describes the evolution of the population state. Our model describes not only the task allocation among the robots, but also the distribution of the robots over the operating space [56]. Such model can be used to predict the evolution of the population and, subsequently, design controllers or supervisors capable of changing the population behavior by the suitable adjustment of appropriate control parameters.

The state-of-the-art in multi-robot system control research concerns research on distributed robot control. It has been shown that distributed control in Robotics and distributed decision making in social insects leads to robust responses to environmental conditions or the population size. However, this work concerns the centralized control of robotic populations, as we claim that centralized control is important for achieving a controllable composition of the population behavior.

The meaning of central control of population is that the human operator, or centralized controller, can send commands for task execution, task cancellation or task switching. This way the population is controllable as a single conventional general purpose robot. The execution of a primitive task sequence ensures that the population can accomplish more complex tasks. The centralized control strategy for the large-size robotic population must take into account the uncertainty of each individual robot reaction, due to communication problems, local

D. Milutinovic and P. Lima: Cells & Robots, STAR 32, pp. 67–89, 2007.
springerlink.com

characteristics of the surrounding environment, etc. This uncertainty is included in our modeling approach and the PDE system that describes the state evolution of a robotic population. The proposed centralized controller is based on the Pontryagin-Hamiltonian optimal control theory for PDEs [21, 46], and provides control of the population space distribution shape.

To motivate our modeling approach to a large size robotic population, we introduce an example in Section 7.1. This example shows a scenario where the robots in the population are receiving commands to change their behavior, but their response is a stochastic process. In Section 7.2, we illustrate the capability of our modeling approach to predict the robotic population position PDF. Application of our modeling approach to the robotic population control problem is discussed in Section 7.3, where we assume that the stochastic parameters of robots are control inputs to the robotic population. We introduce the optimal control problem which aims at maximizing the presence of a robotic population in a given region. We apply Minimum Principle for PDE to this problem and present the developed theory in Section 7.4 [55].

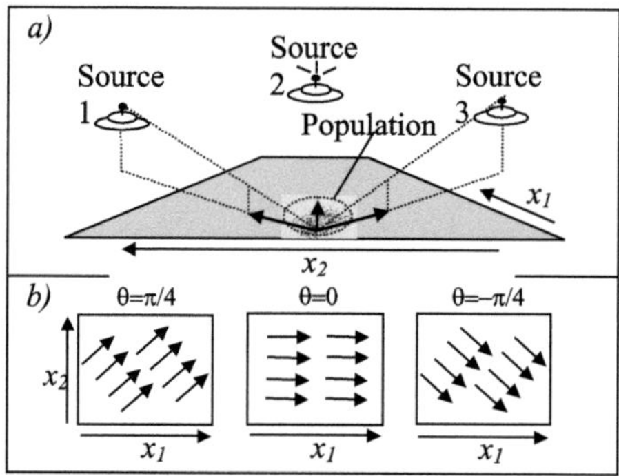

Fig. 7.1. *a*) Robotic population controlled by three aerial robots (sources), *b*) the vector fields created by control signal sources [56], © 2003 IEEE

7.1 Robotic Population Mission Scenario

The scenario we use here introduces the modeling and control approach which assumes a robotic population of *large number* of small mobile robots. The term *small* is applied here to explain that the robot dimension is significantly small compared to the dimensions of the region where the robot is operating. We also assume that the robotic population is sparse and, therefore, no local interaction among the robots is considered.

The robots are initially concentrated on a location over an unexplored terrain (e.g., a mission command station) and controlled by the signals produced by signal sources, e.g., aerial robots (Figure 7.1). The active signal of each source is a command to the robots to change behavior, which in our case means to change the motion direction. Under this scenario, each robot in the population moves in the direction of an active signal source. Under the assumption that the signal sources are far away from the population and that the velocity of the robot is a constant unit ($v = 1$), the robotic motion model is:

$$\dot{x}_1(t) = cos\theta(t) \tag{7.1}$$
$$\dot{x}_2(t) = sin\theta(t) \tag{7.2}$$

where θ is an angle of the motion produced by the signals. In our scenario, we have three signal sources, which means three possible angles of motion, i.e., three possible *discrete states* of the robot. Each discrete state corresponds to one signal source, i.e., one value of motion angle θ. Considering the plane $x_1 \times x_2$ (Figure 7.1b), we assume that Source 1 produces motion with angle $\theta = \pi/4$, Source 2 straight motion $\theta = 0$ and Source 3 produces motion with angle $\theta = -\pi/4$.

There are good reasons for considering that the commands sent to the robots in the population produce the stochastic change of their discrete state, i.e., change of their motion direction. *The common denominator for these reasons is that each robot in the population we are considering is an independent device with its own task priorities.* Because of that, every robot will not react immediately to the command it receives. The actual reaction of the robot to the command depends on many *uncertain factors* such as terrain obstacles, communication visibility, etc., and these factors can be roughly divided into the two classes of reasons:

- physical constraints to the robot;
- limitation of robot resources.

The first class contains factors like the existence of a physical barrier between the specific robot and the signal source, or an obstacle in the direction of the commanded motion. In the latter case, the reaction does not happen even if the command is received correctly.

The second class is related to limited robotic resources. We would like to steer the whole population of robots with an external command. However, not all of the robots will react in the same time. For example, the robot endowed with solar-light re-chargeable batteries might rest till its batteries have been fully re-charged, and would not react to commands meanwhile.

To model the robot which changes the discrete states stochastically, we can use the Micro-Agent based modeling approach. In our examples, we assume that stochastic transitions over different robotic behaviors, i.e., transitions over discrete states, can be modeled as a continuous time Markov Chain. Because of this assumption, we can use the $CTMC\mu A$ depicted in Figure 7.2 to model the robotic population of our example.

The discrete state of the robot depends on the active signal source which forces the robot to move under the vector field $f_i(x) = [cos(\theta_i)\ sin(\theta_i)]^T$; in this example $\theta_1 = -\pi/4$, $\theta_2 = 0$, $\theta_3 = \pi/4$. The stochastic transition from the discrete state i to the discrete state j is described by the transition rate λ_{ij},

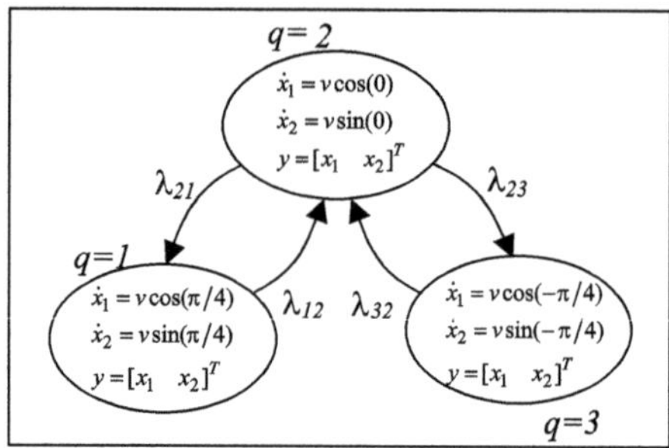

Fig. 7.2. Stochastic Micro-Agent model of an individual robot. The robot can be in one of the discrete states q. In each state, it has different motion defined by the state variable $x = [x_1\ x_2]$. Position $y = [x_1\ x_2]$ is the output of the model. The input is the stochastic sequence of events, which generates the transitions from the state i to the state j, λ_{ij} [56], © 2003 IEEE.

$i, j = 1,\ 2,\ 3$. Therefore, if we define the vector of the Micro-Agent discrete state probabilities $P = [P_1\ P_2\ P_3]^T$, the evolution of that vector obeys

$$\dot{P} = L^T P(t) \tag{7.3}$$

where the Markov Chain transition matrix L is given by:

$$L^T = \begin{bmatrix} -\lambda_{12} & \lambda_{21} & 0 \\ \lambda_{12} & -\lambda_{21} - \lambda_{23} & \lambda_{32} \\ 0 & \lambda_{23} & \lambda_{32} \end{bmatrix} \tag{7.4}$$

The $CTMC\mu A$ state PDF is a vector of the three functions:

$$\rho(x,t) = [\rho_1(x,t)\ \rho_2(x,t)\ \rho_3(x,t)]^T \tag{7.5}$$

where $x = [x_1\ x_2]^T$ is the vector of the robot position. The vector element $\rho_i(x,t)$, $i = 1,2,3$ is the PDF of the robots in the discrete state i at the position x. The time evolution of the $CTMC\mu A$ state PDF satisfies the following system of PDEs:

$$\begin{bmatrix} \rho_1(x,t) \\ \rho_2(x,t) \\ \rho_3(x,t) \end{bmatrix} = L^T \begin{bmatrix} \rho_1(x,t) \\ \rho_2(x,t) \\ \rho_3(x,t) \end{bmatrix} - \begin{bmatrix} \nabla \cdot (f_1(x)\rho_1(x,t)) \\ \nabla \cdot (f_2(x)\rho_2(x,t)) \\ \nabla \cdot (f_3(x)\rho_3(x,t)) \end{bmatrix} \tag{7.6}$$

The PDF of the robot position $\eta(x,t)$ at position x and time t is given by the "total mass" of the probability given x and t, i.e, it is the sum of state PDF components $\rho_i(x,t)$

$$\eta(x,t) = \rho_1(x,t) + \rho_2(x,t) + \rho_3(x,t) \tag{7.7}$$

The equations (7.6) and (7.7) give us the capability to predict the time evolution of the robots position PDF.

The robotic scenario we introduce here is the simplified version of possible real-world examples. One major simplification is that the communication among the robots is not considered. This helps us achieve a mathematical description suitable to model large-size robotic populations which can be exploited for model-based control design. The local coordination among the robots may be included in the model through the vector fields f_i or the transition matrix L that can depend on ρ and its derivatives, i.e., $\frac{\partial \rho}{\partial t}$. Taking into account that L depends on the position x, different environmental conditions in the operating region can be included. For example, in the rocky part of the operating region we would expect that the probability of changing the motion direction would be considerably higher than on the flat part of the terrain.

7.2 Robotic Population Position Prediction

To illustrate our modeling approach to the robotic population of the previous section, we predict here the evolution of the robotic population position PDF. In order to predict the position PDF evolution, we should know the initial PDF and the discrete state transition rates. By using these data, we can generate the position PDF prediction by solving equation (7.6) and, then, applying equation (7.7).

We consider below the prediction of robots position PDF for the model depicted in Figure 7.2 and the two sets of transition rate parameters. The vector of the initial state PDFs is

$$\rho(x,0) = \begin{bmatrix} \rho_1(x,0) \\ \rho_2(x,0) \\ \rho_3(x,0) \end{bmatrix} = \begin{bmatrix} 0 \\ N\left(0_{2\times1}, diag(\sigma^2, \sigma^2)\right) \\ 0 \end{bmatrix} \tag{7.8}$$

which means that at the time $t = 0$ all of the robots are moving straight and their distribution is 2D Gaussian $N\left(0_{2\times1}, diag(\sigma^2, \sigma^2)\right)$, with zero mean and diagonal covariance matrix with $\sigma = 0.1$. This choice of the initial condition means that the initial discrete state probabilities are given by

$$P(0) = [P_1(0)\ P_2(0)\ P_3(0)]^T = [0\ 1\ 0]^T \tag{7.9}$$

To solve the PDE system (7.6), we need boundary conditions. They should be carefully chosen in order to avoid a conflict between the PDE solution and the boundary condition constraint. The useful direction to solve this conflict is to understand the physics of the problem which is described by the PDE [20]. In our case, we are using the boundary condition

$$\rho(x,t) = [0\ 0\ 0]^T,\ \forall t \in [0,\ T],\ x \in \partial X \tag{7.10}$$

where set ∂X refers to the points on the boundary of the region in which we are solving our PDE system and T denotes the terminal time of our solution.

This condition make sense if the set ∂X region cannot be reached in the state space during the time interval $[0, T]$. In our case, the terminal time T is 2h. The transition parameters used for the prediction of the position PDF in two cases (I) and (II) are given in Table 7.1

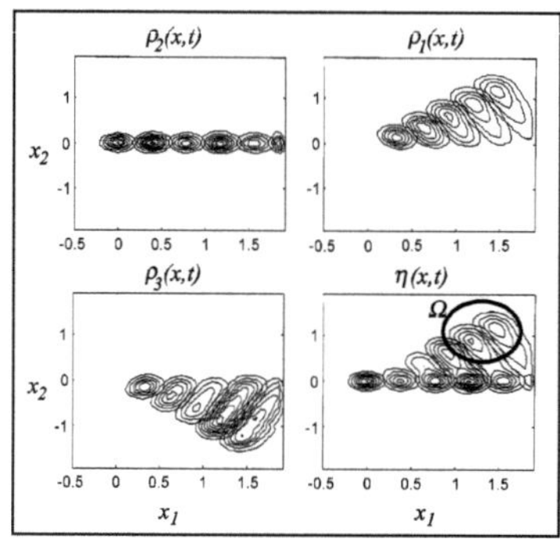

Fig. 7.3. PDF of the robot population states $\rho_i(x, t)$, the PDF of the robots position $\eta(x, t)$, the transition rates Case I, Table 7.1, the plots shown at time instants $t = 0, 0.39, 0.79, 1.18, 1.57, 1.96h$ (from the left to the right in the picture) [56], © 2003 IEEE

Table 7.1. $CTMC\mu A$ model parameter values $[h^{-1}]$

Case	λ_{12}	λ_{21}	λ_{23}	λ_{32}
I	0.1	0.9	0.1	0.5
II	0.4	0.5	0.5	0.1

The evolution of the robotic population state and the position PDF in Case I are presented in Figure 7.3. As it can be seen in the figure, the population dominantly moves in the direction of the vector field $f_1(x)$. The PDF of discrete state 2, $\rho_2(x, t)$ has some initial influence on the shape of the position PDF $\eta(x, t)$ at the beginning of the observed time interval, but then this influence diminishes. Since the contour shape of $\rho_3(x, t)$ has little influence on the shape of the contour plot $\eta(x, t)$, we can conclude that the influence of vector field $f_3(x)$ on the population position PDF shape is insignificant. Both of these conclusions are reasonable and also come from the comparison of the transition rates.

In Case II, the state PDF of the $CTMC\mu A$ states over the discrete space is more uniform. This explains why $\eta(x, t)$ has a more symmetrical shape than in Case I. However, it could be noticed from the density of the contours that the probability of the $CTMC\mu A$ remaining in state 3 is slightly higher than any other. The analysis of the transition rate values shows that this is due to the discrete state probability $P_3(t)$ being slightly higher than $P_1(t)$ and $P_2(t)$.

From the previous results, we may conclude that, in these examples, the dominant direction of the population motion is directed by the vector field $f_i(x)$ of the most probable discrete state i. However, the population spreading shape depends in general on the initial PDF and transition matrix L.

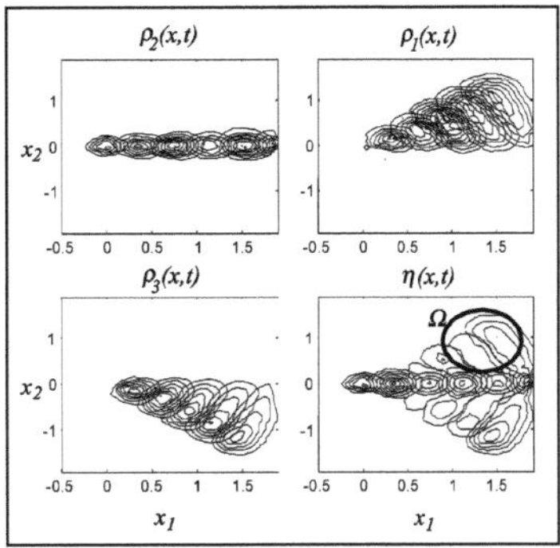

Fig. 7.4. PDF of the robots population states $\rho_i(x,t)$, the PDF of the robots position $\eta(x,t)$, the transition rates Case II, Table 7.1, the plots shown at time instants $t = 0,\ 0.39,\ 0.79,\ 1.18,\ 1.57,\ 1.96h$ (from the left to the right in the picture) [56], © 2003 IEEE

The capability to predict the influence of random changes of robotic behavior in the population and the population position PDF is interesting for the purpose of analysis, but also for the design of control algorithms for large robotic populations.

7.3 Robotic Population Optimal Control Problem

Considering the results of the previous section, we will now introduce an optimal control problem for a large-size robotic population [55]. Let us suppose that the region $\Omega \in X$ in Figure 7.3 and 7.4 is of some particular interest for our robotic mission task and that we want a maximum possible number of robots at time instant T in that region. It is obvious that the set of the transition rates in Case I, Table 7.1, satisfied better our preference than the set of the transition rates in Case II, Table 7.1. Assuming that the transition rate matrix depends on a control vector $u(t) = [u_1(t)\ u_2(t)\ u_3(t)\ u_4(t)]^T$ as $\lambda_{12}(t) = u_1(t)$, $\lambda_{21}(t) = u_2(t)$, $\lambda_{23}(t) = u_3(t)$, $\lambda_{32}(t) = u_4(t)$, we can formulate the optimal control problem to find a control $u(t)$ such that the probability of robotic position $\eta(x,T)$ is maximal, i.e., the following cost function is minimal:

$$J(u) = -\int_{\Omega} \eta(x,T)dx = -\int_{\Omega} (\rho_1(x,T) + \rho_2(x,T) + \rho_3(x,T))dx \qquad (7.11)$$

In this optimal control problem, the transition rates are the result of the environment and limited robot resources. We assume that we can control these rates. For example, if the commands to the population are re-sent with some frequency, we can control the transition rates by controlling this frequency. A higher re-sending frequency increases the probability that the robot will receive the command correctly. This is an open loop control problem and we do not take into account how the changes of transition rates due to uncertain environment changes should be compensated.

Criterion (7.11) can be rewritten in a more general form as:

$$J(u) = -\int_X w^T(x)\rho(x,T)dx \tag{7.12}$$

where u is a vector of the population parameters that may be set externally by an appropriate command sent to the population and $w(x) = [w_1(x)\ w_2(x)\ w_3(x)]$ is a vector of weighting functions:

$$w_i(x) = \begin{cases} 1, & x \in \Omega \\ 0, & otherwise \end{cases} \quad , i = 1,2,3 \tag{7.13}$$

The vector of weighting functions is introduced because it allows us to specify the other types of optimal criterion we can minimize. Two other examples of weighting functions are:

$$w_i(x) = \begin{cases} 1, x \in \Omega_1 \cup \Omega_2, \Omega_1, \Omega_2 \in \Omega \\ 0, \qquad\quad otherwise \end{cases} \tag{7.14}$$

$$w_i(x) = \begin{cases} e^{(x-x)^T \Sigma^{-1}(x-x)}, & x \in \Omega \\ 0, \qquad\quad otherwise \end{cases} \tag{7.15}$$

Using the weighting (7.14) in criterion (7.12) means the maximization of the probability over the region composed of two parts which are not necessarily connected. Obviously the value of the weighting function can be different from either 0 or 1. Using a weighting function such as (7.15) also defines a preferred PDF shape in the region Ω.

In the text below, we will assume that our robotic population can be modeled by a $CTMC\mu A$ which can be controlled and its state PDF evolution is given by the following PDE system:

$$\frac{\partial \rho(x,t)}{\partial t} = L^T(u)\rho(x,t) - \begin{bmatrix} \nabla \cdot (f_1(x)\rho_1(x,t)) \\ \nabla \cdot (f_2(x)\rho_2(x,t)) \\ \vdots \\ \nabla \cdot (f_N(x)\rho_N(x,t)) \end{bmatrix} \tag{7.16}$$

where u is a control input which is a time function

$$u : [0,T] \to U \text{ and } u \in U_{ad}(0,T;U) \tag{7.17}$$

influencing the transition matrix $L(u)$, where $U_{ad}(0, T; U)$ is a set of admissible control functions taking values in the set U. We also assume that the initial condition $\rho(x, 0)$ is given, as well as the boundary condition $\rho(x, t) = 0$, $\forall t \in [0, T]$, $\forall x \in \partial X$, where ∂X is the set of boundary points of the region X. This boundary condition says that the boundary region cannot be reached in the state space by any means during the time interval $[0, T]$, which is reasonable for conveniently chosen vector fields describing the continuous dynamics of each discrete state.

The system of PDEs (7.16) can be written as:

$$\frac{\partial \rho(x, t)}{\partial t} = F(u)\rho(x, t) \tag{7.18}$$

where F is a linear operator:

$$F(u) = L^T(u) - diag(\phi_1, \phi_2, \ldots \phi_N) \tag{7.19}$$

$$\phi_k = \sum_{j=1}^{n} \frac{\partial f_k^j(x)}{\partial x_j} + f_k^j(x)\frac{\partial}{\partial x_j}, k = 1, 2, \ldots N \tag{7.20}$$

where $f_i^j(\mathrm{x})$ is the jth component of the vector $f_i(x) = [f_i^1(x)\ f_i^2(x) \ldots f_i^n(x)]$. The index is $j = 1, 2, \ldots n$ since $x \in R^n$.

The optimal control problem can be summarized as to determine the optimal control $u = u^*$ which *minimizes* the objective function:

$$J(u) = -\int_X w(x)^T \rho(x, T)dx \tag{7.21}$$

under the constraint

$$\frac{\partial \rho(x, t)}{\partial t} = F(u)\rho(x, t) \tag{7.22}$$

where $w(x)^T = [w_1(x)\ w_2(x) \ldots w_N]$ is a vector of weighting functions, and T is a time instant when the objective function should be minimal. We should not forget that under this condition, the corresponding initial $\rho(x, 0)$ and boundary $\rho(x, t), x \in \partial X$ conditions must be satisfied.

The optimal control problem (7.16)-(7.21) is a PDE optimal control problem. To solve this problem, the *Minimum Principle for PDE* [21] can be exploited in a similar way as *Pontryagin Minimum Principle* can be exploited to solve ODE optimal control problems [50]. Along this line of reasoning, our control problem can be considered as a special case of a more general optimal control problem of the *evolution equation* (D.1) (see Appendix D), where:

$$\rho'(t) = \frac{\partial \rho(x, t)}{\partial t} \tag{7.23}$$

$$f(\rho(t), u(t), t) = F(u(t))\rho(x, t) \tag{7.24}$$

Considering the criterion (D.2) that must be minimized in the general problem formulation (Appendix D), we can find that

$$f_0(\rho(t, u), u(t), t) = 0 \tag{7.25}$$

$$\rho_0(T, u) = g_0(\rho, \rho(T, u)) = J(u) = \langle -w(x), \rho(x, T) \rangle \tag{7.26}$$

where the scalar product of the two functions, say $\phi(x)$ and $\psi(x)$, is denoted as $\langle \phi, \psi \rangle$ and defined as

$$\langle \phi, \psi \rangle = \int_X \phi^T(x)\psi(x)dx \tag{7.27}$$

Since the operator $F(u)$ is linear and $f_0(\rho(t), u(t), t) = 0$, the Fréchet derivatives of $f(\rho(t), u(t), t)$ and $f_0(\rho(t), u(t), t)$, regarding ρ, are given by:

$$\partial_\rho f(\rho(t), u(t), t) = F(u(t)) , \quad \partial_\rho f_0(\rho(t), u(t), t) = 0 \tag{7.28}$$

Thus, under the condition that the linear $F(u)$ operator is bounded, $\|F(u(t))\| < \infty$, we have satisfied all the conditions (D.11)-(D.14) and we can apply the corresponding *Minimum Principle*, (Theorem D.1 in Appendix D). This theorem ensures the existence of the scalar z_0 and the vector functions $z(x) = [z_1(x) \ z_2(x) \ z_N(x)]^T$ such that the optimal control satisfies:

$$u^* = \arg \min_{u \in U_{ad}} H(\rho^*(x, t), u(t), t) = \arg \min_{u \in U_{ad}} \langle \pi, F\rho^* \rangle \tag{7.29}$$

where $\pi(x, t)$ is the solution of the adjoint equation that depends on z_0 and z:

$$\frac{\partial \pi(t)}{\partial t} = -F^T(u^*)\pi(t) \tag{7.30}$$

$$\pi(T) = z - z_0 w(x) \tag{7.31}$$

and $\rho^*(x, t)$ is the solution of the equation when the optimal control u^* is applied, i.e.:

$$\frac{\partial \rho^*(x, t)}{\partial t} = F(u^*(t))\rho^*(x, t), \ \ \rho^*(x, 0) = \rho(x, 0) \tag{7.32}$$

Since in our problem formulation we do not have a target condition, which must be satisfied at time T, Theorem 4 can be applied with:

$$z_0 = 1, \ z(x) = 0 \tag{7.33}$$

Then the adjoint equation becomes

$$\frac{\partial \pi(t)}{\partial t} = -F^T(u^*)\pi(t) , \quad \pi(T) = -w(x) \tag{7.34}$$

To solve the optimal control problem, we should search for the control u^* which simultaneously satisfies Equations (7.29), (7.32), and (7.34). In ODE optimal control theory this problem is the so-called two-point-boundary-value problem, and in some cases that problem can be handled either analytically or numerically. Solving the PDE optimal control is even more difficult because the "values" at two points are functions. Still, similarly as in the case of ODE optimal control, it is possible that, based on this equation, we can find an optimal control. However, it is also possible that (7.29), (7.32), and (7.34) are in such forms that we cannot infer what the optimal control is. This is a problem of "singular controls" within the optimal control problem which has to be investigated carefully. In the following section, we will present an example that illustrates the previously presented modeling and optimal control theory.

7.4 Example of Using the PDE Minimum Principle for Robotic Population Control

The example presented here is very similar to the one presented in the introductory Section 7.1. However, for the purposes of simplicity, we assume that each robot can move only in one dimension, which means that they can move left, right or stop [55]. Therefore, a one dimensional robots position distribution will be considered. Robots are controlled by three signals L, R and S (Fig. 7.5a). If the signal L is active, it attracts robots to move left. Signal R attracts robots to move right and signal S send a stop command. Obviously, each robot in the population can be either in state $1 = move\_left$, $2 = move\_right$ or $3 = stop$. If the discrete dynamics of these three states can be described by a Markov Chain, the full state of this robot population can be described by a $CTMC\mu A$, see figure 7.5b. The transition rates $u_1(t)$, $u_2(t)$ and $u_3(t)$ are the control variables and we assume that $u_1(t), u_2(t), u_2(t) \in [0, u_{max}]$.

To formulate the control problems, we will assume that:

- the motion to the left is given by

$$\dot{x}(t) = f_1(x,t) = k_1, \; k_1 = 0.5 \tag{7.35}$$

- the motion to the right is given by

$$\dot{x}(t) = f_2(x,t) = k_2, \; k_2 = -0.25 \tag{7.36}$$

- and obviously, the stop motion is given by

$$\dot{x}(t) = 0 \tag{7.37}$$

The initial robotic distribution, i.e., the PDF of the robot positions $\eta(x,0)$, is

$$\eta(x,0) = \rho_1(x,0) + \rho_2(x,0) + \rho_2(x,0) \tag{7.38}$$

where $\rho_1(x,0)$, $\rho_2(x,0)$ and $\rho_3(x,0)$ are the initial PDFs of the robots moving left, moving right and not moving, $\rho_1(x,0) = 0$, $\rho_2(x,0) = 0$ and

$$\rho_3(x) = \begin{cases} \frac{1}{\sqrt{0.02\pi}} exp(-\frac{(x-2.5)^2}{0.02}), & 2 < x < 3 \\ 0, & otherwise \end{cases} \tag{7.39}$$

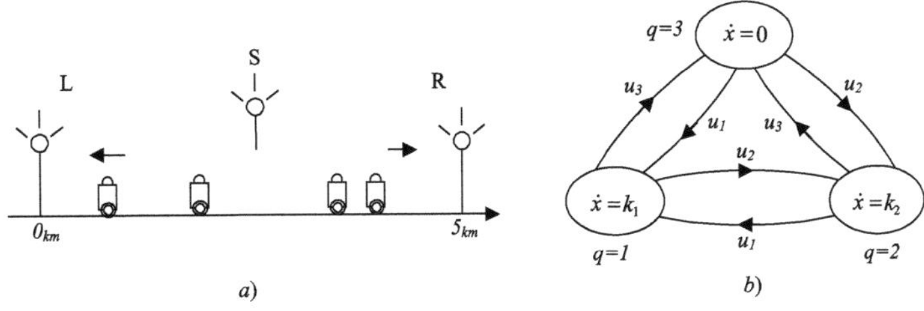

Fig. 7.5. *a)* A robotic population controlled by three signal sources, *b)* The robotic population Micro-Agent model [55], © 2006 IEEE

respectively. Defining the time duration of the robotic mission as $T = 3h$ and the weighting function

$$w_3(x) = \begin{cases} \frac{1}{\sqrt{0.01\pi}} exp(-\frac{(x-1.75)^2}{0.01}), & 1.25 < x < 2.25 \\ 0, & otherwise \end{cases} \tag{7.40}$$

the optimal control problem is to find the optimal transition rates $u_1(t), u_2(t), u_3(t) \in [0, u_{max}]$, $u_{max} = 2$, such that the objective function

$$J = -\int_X \underbrace{[0\ 0\ w_3(x)]}_{w^T} \rho(x,T)dx, \quad X = [x_a, x_b], \ x_a = 0, \ x_b = 5 \tag{7.41}$$

is minimal. The motivation for the weighting function $w_3(x)$ choice is to compute a control that will move the center of the robotic distribution 2.5 to 1.75, and make the distribution slightly sharper.

The following system of equations describes the state PDF evolution of a robotic $CTMC\mu A$

$$\begin{bmatrix} \frac{\partial \rho_1(x,t)}{\partial t} \\[2mm] \frac{\partial \rho_2(x,t)}{\partial t} \\[2mm] \frac{\partial \rho_3(x,t)}{\partial t} \end{bmatrix} = (F_u(u(t)) + F_\partial) \begin{bmatrix} \rho_1(x,t) \\[2mm] \rho_2(x,t) \\[2mm] \rho_3(x,t) \end{bmatrix} \tag{7.42}$$

with

$$F_u(u) = \begin{bmatrix} -u_2 - u_3 & u_1 & u_1 \\[2mm] u_2 & -u_1 - u_3 & u_2 \\[2mm] u_3 & u_3 & -u_1 - u_2 \end{bmatrix} \tag{7.43}$$

$$F_\partial = diag(-k_1\frac{\partial}{\partial x}, \ -k_2\frac{\partial}{\partial x}, \ 0) \tag{7.44}$$

where $u(t) = [u_1(t)\ u_2(t)\ u_3(t)]^T$ and t in $u(t)$ is omitted in (7.43). Equation (7.42) can be taken in the form of equation (7.18), i.e.:

$$F(u(t)) = F_u(u(t)) + F_\partial \tag{7.45}$$

To apply Theorem D.1, in Appendix D, we should first define a state space for ρ and its co-state π. We will assume that the solutions ρ, π of (7.42) and (7.34), respectively, are from the same set of functions E that satisfies:

$$\rho, \ \pi \in E \tag{7.46}$$
$$\rho, \ \pi \ : \ X \rightarrow L^2 \times L^2, \partial X \rightarrow 0 \tag{7.47}$$
$$d(\rho_i^2), \ d(\pi_i^2) \ : \ \text{Lebesgue integrable}, \ i = 1,2 \tag{7.48}$$

The symbol L^2 stands for the set of Lebesgue measurable functions (see Appendix D) and symbol d for the differentiation. Therefore, we take $x_a = 0$, $x_b = 5$ in $X = [x_a, x_b]$ and assume the boundary conditions

$$\rho(0, t) = \rho(5, t) = \pi(0, t) = \pi(5, t), t \in [0, T]. \tag{7.49}$$

The *nonzero* intervals of $w_3(x)$ and $\rho_3(x, 0)$ ensure that these boundary conditions are satisfied, i.e., due to the finite movement speed there is not a single robot from the population that has time to reach the space X boundary position x_a or x_b.

Using the scalar product defined by (7.27), the space E is a *Hilbert* space with the norm:

$$\|\rho\|_E = \langle \rho, \rho \rangle^{\frac{1}{2}} \tag{7.50}$$

Therefore, from (7.24), we have

$$\|f(x, \rho, u)\| = \|F(u)\rho\| = \langle F(u)\rho, F(u)\rho \rangle \tag{7.51}$$

and the operator norm

$$\|F(u)\| = \max_{\|\rho\| \leq 1} \|F(u)\rho\| \tag{7.52}$$

To show that the operator $F(u)$ is bounded, we use the triangular inequality:

$$\|F(u)\| = \|F_u(u) + F_\partial\| \leq \|F_u(u)\| + \|F_\partial\| \tag{7.53}$$

The linear operator F_∂ is symmetric and we will use it to find its norm as:

$$\|F_\partial\| = \max_{\|\rho\| \leq 1} |\langle \rho, F_\partial \rho \rangle| \tag{7.54}$$

Applying the definition of scalar product (7.27), we have

$$|\langle \rho, F_\partial \rho \rangle| = \left| \int_{x_a=0}^{x_b=5} -k_1 \rho_1 \frac{\partial \rho_1}{\partial x} dx + \int_{x_a=0}^{x_b=5} -k_2 \rho_2 \frac{\partial \rho_2}{\partial x} dx \right| \tag{7.55}$$

where the term with k_3 does not appear, since $k_3 = 0$. By virtue of the space E property (7.47):

$$\int_0^5 \frac{1}{2} \rho_i \frac{\partial \rho_i}{\partial x} = \int_0^5 d(\rho_i^2) = \rho_i^2(x_b, t) - \rho_i^2(x_a, t) = 0, \ i = 1, 2 \tag{7.56}$$

$$|\langle \rho, F_\partial \rho \rangle| = 0, \forall \rho \tag{7.57}$$

we can conclude that

$$\|F_\partial\| = 0 \tag{7.58}$$

$F_u(u)$ is a matrix with finite real coefficients corresponding to the values in the admissible set of the control U_{ad}. For any choice of admissible control values, there exists a maximal singular value $\sigma_{\max}(F_u(u))$, which is a finite number. Therefore we can draw out the conclusion that the operator $F(u)$ is bounded, i.e.:

$$\|F(u)\| \leq \max_{u_1, u_2, u_3 \in U_{ad}} \|F_u(u)\| \tag{7.59}$$

and we can exploit Theorem D.1, in Appendix D, to find the optimal control.

According to (7.34) and noting that:

$$F_{\partial}^{T} = -F_{\partial} \tag{7.60}$$

the system of equations which describes the evolution of the co-state is:

$$\begin{bmatrix} \frac{\partial \pi_1(x,t)}{\partial t} \\ \frac{\partial \pi_2(x,t)}{\partial t} \\ \frac{\partial \pi_3(x,t)}{\partial t} \end{bmatrix} = -(F_u^T(u^*(t)) - F_{\partial}) \begin{bmatrix} \pi_1(x,t) \\ \pi_2(x,t) \\ \pi_3(x,t) \end{bmatrix} \tag{7.61}$$

$$\pi_1(x,t) = \pi_2(x,t) = 0, \ \pi_3(x,t) = -w_3(T) \tag{7.62}$$

and the optimal control $u^*(t) = [u_1^*(t) \ u_2^*(t) \ u_3^*(t)]^T$ satisfies:

$$u^*(t) = \arg \min_{u \in U_{ad}} H(t, \rho^*, u) \tag{7.63}$$

Using the definition of scalar product (7.27)

$$u^*(t) = \arg \min_{u \in U_{ad}} \int_X \pi(t)(F_u(u) + F_{\partial})\rho^*(t)dx \tag{7.64}$$

Since in this expression only F_u depends on u, we have

$$u^* = \arg \min_{u \in U_{ad}} \int_{x_a=0}^{x_b=5} \pi(t)(F_u(u))\rho^* dx \tag{7.65}$$

which can be presented in a linear form as

$$u^* = \arg \min_{u \in U_{ad}} [u_1(t)I_1(t) + u_2 I_2(t) + u_3(t)I_3(t)] \tag{7.66}$$

where

$$I_1(t) = \int_0^5 (\pi_1 - \pi_2)\rho_2^* + (\pi_1 - \pi_3)\rho_3^* dx$$

$$I_2(t) = \int_0^5 (\pi_2 - \pi_1)\rho_1^* + (\pi_2 - \pi_3)\rho_3^* dx \tag{7.67}$$

$$I_3(t) = \int_0^5 (\pi_3 - \pi_1)\rho_1^* + (\pi_3 - \pi_2)\rho_2^* dx$$

and $\rho_i^* = \rho_i^*(x,t)$, $\pi_i = \pi_i(x,t)$ are computed for u^*. If we can compute $I_1(t)$, $I_2(t)$ and $I_3(t)$, then the optimal control $u^*(t)$ will be defined as follows:

$$u_i^* = \begin{cases} 0, & I_i(t) > 0 \\ u_i^{max}, & I_i(t) < 0 \ , \ i = 1,2,3 \\ u_i^* \in U_{ad}, & I_i(t) = 0 \end{cases} \tag{7.68}$$

To compute the optimal control using this expression, we should prove that either $I_i(t) \neq 0$, $\forall t[0,\ T]$, or that $I_i(t) = 0$ only for a discrete set of time instants t_k. Otherwise, we can conclude that expression (7.66) is still valid, but cannot help us calculate the control $u_i(t)$, because for $I_i(t) = 0$, the control can take any value from the admissible set U_{ad}. This is the "singular control" problem. In that case some further structural properties of PDE systems (7.42) and (7.61) must be examined [55].

One way to deal with the "singular control" problem is to modify the cost function (7.41) adding the term which depends on control u and is penalized by parameter $\varepsilon > 0$. This results in [55]

$$J^\varepsilon = -\int_X w^T \rho(x, T) + \varepsilon \int_0^T u_1^2(t) + u_2^2(t) + u_3^2(t)dt \qquad (7.69)$$

Using a small ε, we have $J \approx J^\varepsilon$. Applying the Minimum Principle, the optimal control u^* should satisfy

$$u^*(t) = \arg \min_{u \in U_{ad}} \int_X \pi(t)(F_u(u) + F_\partial)\rho^*(t) + \varepsilon u^T u \qquad (7.70)$$

that, similarly as in equation (7.66), results in

$$u^* = \arg \min_{u \in U_{ad}} \left[u_1(t)I_1(t) + u_2I_2(t) + u_3(t)I_3(t) + \varepsilon(u_1^2 + u_2^2 + u_3^2) \right] \qquad (7.71)$$

i.e.,

$$u^* = \arg \min_{u \in U_{ad}} H^\varepsilon(t, \rho^*, u) \qquad (7.72)$$

since the Hamiltonian H^ε, corresponding to the problem with the cost function J^ε, is

$$H^\varepsilon(t) = H(t) + \varepsilon(u_1^2(t) + u_2^2(t) + u_3^2(t)) \qquad (7.73)$$

The first partial derivative of this Hamiltonian regarding the control component $u_i(t)$ is

$$\frac{\partial H^\varepsilon(t)}{\partial u_i(t)} = I_i(t) + 2\varepsilon u_i(t) \qquad (7.74)$$

and solving $\frac{\partial H^\varepsilon(t)}{\partial u_i(t)} = 0$, we find that the optimal control can be expressed as

$$u_i^* = \begin{cases} 0, & -\frac{I_i(t)}{2\varepsilon} < 0 \\ u_i^{max}, & -\frac{I_i(t)}{2\varepsilon} > u_{max}, \ i = 1, 2, 3 \\ -\frac{I_i(t)}{2\varepsilon}, & elsewhere \end{cases} \qquad (7.75)$$

Now, the problem of the singular control does not exist. The above expression is exact, but u_i^* can be computed only when I_i is known at each point. Since we do not have an expression for I_i, and it also depends on u^*, the only way to compute the optimal control is to apply some iterative numerical procedure that will compute u^*, which simultaneously satisfies (7.42), (7.61), (7.67) and (7.72). Avoiding the "singular control" problem presented here results in an iterative numerical algorithm for computing the optimal control, proposed in the next section, that will not indefinitely stop at the points where $I_i(t) = 0$, without warranty that the computed control is optimal. The price paid is that we are not solving the original control problem, but the control problem with the cost function J^ε.

7.4.1 Complexity of Numerical Optimal Control

To solve the optimal control problem of the previous section numerically, we can deal with the discrete approximations of the control $u(t)$ and the functions $\rho(x,t)$, $\pi(x,t)$ and $w(x)$. This also results in dealing with the approximative values of the cost function $\hat{J}^\varepsilon$ and the corresponding Hamiltonian $\hat{H}^\varepsilon$. The control computed in this way is so-called numerical optimal control [78].

The time discretization of control $u(t)$ results in

$$\hat{u}(k) = u(k\Delta T), \ \ k = 0, 1, 2, ... K - 1 \tag{7.76}$$

and we can use it to approximate the control as

$$u(t) \approx \hat{u}(k), \ \ k\Delta T < t < (k+1)\Delta T, \tag{7.77}$$

where $\hat{u}(k)$ are the values of the sequence $\hat{u}$ approximating control and ΔT is the discretization time step. If T is the interval of time for which we compute the optimal control, then the sequence $\hat{u}$ has the finite length $K = T/\Delta T$. Similarly, the discretization of $\rho(x,t)$ and $w(x)$ results in

$$\hat{\rho}(m, k) = \rho(m\Delta X, k\Delta T), \ \ \ k = 0, 1, 2, \dots K \tag{7.78}$$
$$\hat{\pi}(m, k) = \pi(m\Delta X, k\Delta T), \ \ \ m = 1, 2, \dots M \tag{7.79}$$
$$\hat{w}(m) = w(m\Delta X) \tag{7.80}$$

with ΔX, which is the space discretization step. In our case, $X = [x_a, x_b]$ and, naturally, $M = 1 + (x_b - x_a)/\Delta X$. From the control vectors $\hat{u}(k)$, we can build the $3K \times 1$ vector $\hat{u}$. Similarly, from the vectors $\hat{\rho}(m, k)$ and $\hat{\pi}(m, k)$, we can build the $3M \times 1$ vectors $\hat{\rho}(k)$ and $\hat{\pi}(k)$. In this and the next section, we will use the vectors $\hat{u}$, $\hat{\rho}(k)$ and $\hat{\pi}(k)$ whenever it is convenient, in order to simplify the notation.

The numerical optimal control problem is to find the optimal control sequence values $\hat{u}^*(k) = [u_1^*(k) \ u_2^*(k) \ u_3^*(k)]^T$ which minimize the cost function $\hat{J}^\varepsilon$. This cost function is the cost function J^ε computed based on the discrete approximations listed above. The optimal control problem can be written as

$$\hat{u}^* = \arg \min_{\hat{u} \in U_{ad}} \hat{J}^\varepsilon(\hat{u}) \tag{7.81}$$

and one can try to solve this problem using some standard iterative numerical minimization algorithm [8]. Usually, in each iteration j, the algorithm applies some update rule producing the "better" control $\hat{u}^j$, such that the cost function based on this control $\hat{J}^\varepsilon(\hat{u}^j)$ is smaller then the cost function from the previous iteration $j-1$, i.e., $\hat{J}^\varepsilon(\hat{u}^j) < \hat{J}\varepsilon(\hat{u}^{j-1})$. We say usually, because the update rule can also choose another direction for the control update in order to increase the chance of reaching the minimum described by the equation (7.81). In any case, the iterative procedures always produce the sequence of control $\hat{u}^j$, which converges to $\hat{u}^*$, such that $\hat{u}^* = \arg\min_{\hat{u}(k)\in U_{ad}} \hat{J}^\varepsilon(\hat{u}^*)$

Let us assume that the iterative procedure only needs the first order derivatives of $\hat{J}^\varepsilon$, i.e., its gradient $\nabla_{\hat{u}}\hat{J}^\varepsilon$. To compute the derivatives $\partial\hat{J}^\varepsilon(\hat{u})/\partial\hat{u}(0)$, we need to know the value of $\hat{J}^\varepsilon(\hat{u})$. For computing this value, we need to compute $\hat{\rho}(m,K)$, $\forall m = 1,\dots M$, which can be computed only when $\hat{\rho}(m, K-1)$, $\forall m$ is known; but to obtain this, we need the value of $\hat{\rho}(m, K-2)$ $\forall m$, etc. To obtain the value of $\hat{J}(\hat{u})$ in our example, we have to evaluate $K \cdot 3M \cdot N$ values. The same number of values must be evaluated for $\hat{J}^\varepsilon$ when each of the three components of $\hat{u}(0) = [\hat{u}_1(0)\ \hat{u}_2(0)\ \hat{u}_3(0)]^T$ is perturbed. Therefore, for computing $\partial\hat{J}^\varepsilon(\hat{u})/\partial\hat{u}(0)$, it is necessary to evaluate $3(K+1)NM$ values. For any additional derivative $\partial\hat{J}^\varepsilon(\hat{u})/\partial\hat{u}(k)$, it is necessary to evaluate additional $3(K-k)NM$ values. In summary, to evaluate all the components of the gradient $\nabla_{\hat{u}}\hat{J}^\varepsilon$, of the cost function $\hat{J}^\varepsilon$, the number of evaluated values C_p is

$$C_p = 3\left(1 + \frac{K(K+1)}{2}\right) NM \tag{7.82}$$

The number C_p can be used as a measure of the computational complexity of the numerical algorithm, since computing $\rho(m,k)$ resulting from the PDE system brings a major computational load. To provide a realistic approximation of the control $u(t)$, the time step ΔT should be small. In other words, the sequence length K should be large. The numerical complexity C_p of the presented approach, in the simplest form, scales with K^2 and it is the major obstacle for its implementation.

Exploiting the Hamiltonian for solving the optimal control problem can reduce the computational complexity. The simplest way to do that is to use iterations that minimize the Hamiltonian $\hat{H}^\varepsilon(k)$ at each time point, i.e., for each k. In other words, for a given control sequence in the iteration j, $\hat{u}^j(k)$, we should perform all the necessary computation to obtain $\hat{H}^\varepsilon(u^j(k))$, for each k. Then, we update the value of control $\hat{u}^j(k)$, so that the updated value $\hat{u}^{j+1}(k)$ minimizes $\hat{H}^\varepsilon(u^j(k))$. We expect that the vector of control $\hat{u}^j$ converges to $\hat{u}^*$, such that the corresponding control sequence satisfies

$$\hat{u}^*(k) = \arg\min_{u(\hat{k})\in U_{ad}} \hat{H}^\varepsilon(\hat{u}^*(k)), \ \forall k \tag{7.83}$$

For computing the Hamiltonian $\hat{H}^\varepsilon(u^j(k))$ for each $k = 0, 1, \dots K$, it is necessary to compute $\rho(m,1), \dots \rho(m,K)$ and $\pi(m, K-1), \dots, \pi(m,1)$, $\forall\ m = 1, \dots M$. This means that we need to evaluate $2K \cdot N \cdot M$ values. The computational complexity of this Hamiltonian-based approach scales with K.

We use this approach in [55] with the discrete time step $\Delta T = 0.03$, $K = 100$, $M = 500$ and the parameter $\varepsilon = 10^-7$. The chosen value for ε depends on how

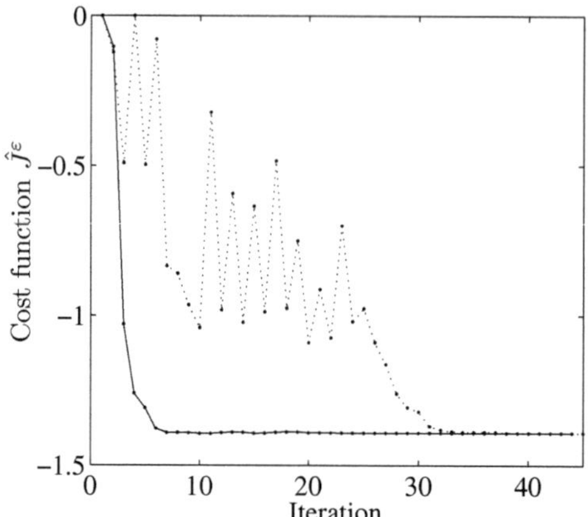

Fig. 7.6. Cost function $\hat{J}^{\varepsilon}$ computed in each iteration of the algorithm (solid) and computed by the simpler algorithm [55], © 2006 IEEE

close to J we want J^{ε} to be; the smaller the ε, the closer J^{ε} to J. For the decimal precision of d_p, the term which penalizes the control in J^{ε} must be smaller than 10^{-d_p}. Since we know the maximal values of control u_{max}, this can be expressed by [55]:

$$10^{-d_p} > J - J^{\varepsilon} = \varepsilon \int_0^T \sum_{i=1}^{3} u_{max}^2 dt = 3T u_{max}^2 \Rightarrow \varepsilon < \frac{10^{-d_p}}{3T u_{max}^2} \qquad (7.84)$$

However, ε cannot be infinitely small because the algorithm convergence may be influenced when the minimization problem is close to the original problem with the "singular control".

For the initial guess for the optimal control in [55], we use $\hat{u}^1(k) = [0.5\ 0.5\ 0.5]$ $\forall k$. To illustrate the convergence, in each iteration j, we also compute the value of the cost function $\hat{J}^{\varepsilon}$. These values are shown in figure 7.6 using a dotted line. In this example, the algorithm reaches the optimal control after 40 iterations. We are not illustrating the shape of the computed optimal control at the moment. It is because the following section describes another algorithm that also has a linear scaling of the complexity C_p with K, and, for the same example, it converges to the same optimal control much faster.

7.4.2 Numerical Optimal Control

We search for the optimal control $\hat{u}^*$ that is the stationary point of the iteration

$$\hat{u}^{j+1} = \hat{u}^j + \alpha^j d^j \qquad (7.85)$$

and $\hat{u} \in U_{ad}$. The variable α^j is the scalar for which $\hat{J}^{\varepsilon}(u^j + \alpha^j d^j) < \hat{J}^{\varepsilon}(u^j)$ and d^j is the search vector in the jth iteration. This vector, for example, can be the negative gradient of the cost function, i.e., $d^j = -\nabla \hat{J}^{\varepsilon}(u^j)$. The gradient vector dimension is $\dim(\nabla \hat{J}^\varepsilon) = \dim(d^j) = \dim(\hat{u}^j) = 3K$. In fact, after the discretization, we deal with the discrete time optimal control problem. Therefore, this gradient can be completely computed using the Hamiltonian values due to the relation $\nabla_{\hat{u}(k)}\hat{J}^\varepsilon = \nabla_{\hat{u}(k)}\hat{H}^\varepsilon$ for the given point at time k [8]. The dimension of these derivatives is $\dim(\nabla_{\hat{u}(k)}\hat{J}^\varepsilon) = \dim(\nabla_{\hat{u}(k)}\hat{H}^\varepsilon = 3)$. On the other side, search for the value of α^j along the direction d^j is computationally expensive, since the cost $\hat{J}^\varepsilon$ evaluation requires the PDE system solution. Therefore, we find the scalar value α^j as the value, providing that the gradient at the point $\hat{u}^j + \alpha^j d^j$ is orthogonal to the direction d^j, i.e.,

$$[\nabla \hat{J}^\varepsilon(\hat{u}^j + \alpha^j d^j)]^T d^j = 0 \tag{7.86}$$

which, due to the quadratic form of the Hamiltonian $\hat{H}^\varepsilon$, has a closed form solution. The minimization based on the iterations with $d^j = -\nabla \hat{J}^\varepsilon(u^j)$ can have a poor convergence [8]. The numerical algorithm used here is based on Nonlinear Conjugate Gradient Method with the Polak-Ribière for the direction d^j generation [64]. While we have explained the main idea of the algorithm we applied $\hat{u}^j$ and d^j, in the following detail algorithm description, we will use $\hat{u}^j(k) = [u_1^j(k)\ u_2^j(k)\ u_3^j(k)]^T$ and $d^j(k) = [d_1^j(k)\ d_2^j(k)\ d_3^j(k)]^T$, representing the control at time point k in the jth iteration and the corresponding update direction, respectively. The algorithm steps are:

Step 1) Guess the control sequence $\hat{u}^1(k)$, $k = 0, 1, 2, \ldots K - 1$ and set the iteration counter $j = 1$.

Step 2) Compute $\hat{\rho}^j(m, k)$, the discrete approximation of the solution $\rho(x, t)$ in the jth iteration, $k = 0, 1, \ldots, K$, $m = 1, 2, \ldots, M$. This computation is based on the control sequence $\hat{u}^j(k)$, starting from the initial value $\hat{\rho}^j(m, 0)$ forward in time.

Step 3) Compute $\hat{\pi}^j(m, k)$, the discrete approximation of the solution $\pi(x, t)$ in the jth iteration $k = K, K - 1, \ldots, 0$, $m = 1, 2, \ldots, M$. This is computed based on the control sequence $\hat{u}^j(k)$, starting from the initial value $\hat{\pi}^j(m, K) = \hat{w}(m)$ backward in time.

Step 4) Using the values from the previous step, compute the discrete approximations $\hat{I}_1^j(k)$, $\hat{I}_2^j(k)$, $\hat{I}_2^j(k)$ of the integrals composing the Hamiltonian in the jth iteration

$$\hat{H}^\varepsilon(k)^j = \hat{u}_1^j(k)\hat{I}_1^j(k) + \hat{u}_2^j(k)\hat{I}_2^j(k) + \hat{u}_3^j(k)\hat{I}_3^j(k) \tag{7.87}$$

Step 5) Compute the negative gradient $g^j(k) = -\nabla_{\hat{u}^j(k)}\hat{H}^\varepsilon(k)^j$ for each $k = 0, 1, \ldots K - 1$, that is

$$g^j(k) = -\nabla_{\hat{u}^j(k)}\hat{H}^\varepsilon(k)^j = \begin{bmatrix} -\hat{I}_1(k) - 2\varepsilon\hat{u}_1(k) \\ -\hat{I}_2(k) - 2\varepsilon\hat{u}_2(k) \\ -\hat{I}_2(k) - 2\varepsilon\hat{u}_2(k) \end{bmatrix} \tag{7.88}$$

Step 6) If $j = 1$, set the direction $d^1(k) = g^j(k)$ for each $k = 0, 1, \ldots K - 1$ and set the scalar value $\beta^1 = 0$. In all other cases compute

$$\beta^j = \max \left\{ \frac{g^j(k)^T (g^j(k) - g^{j-1}(k))}{g^{j-1}(k)^T g^{j-1}(k)}, 0 \right\} \tag{7.89}$$

also whenever $g^{j-1}(k)^T g^{j-1}(k) = 0$ set the $\beta^j = 0$. Then set the directions

$$d^j(k) = g^j(k) + \beta^j d^{j-1}(k), \quad k = 0, 1, 2, \ldots K - 1 \tag{7.90}$$

Step 7) Compute the discrete approximations $\hat{I}_1^j(k)$, $\hat{I}_2^j(k)$, $\hat{I}_2^j(k)$ of the integrals composing the Hamiltonian H^ε. Using these values, compute α^j using the closed form solution of $[\nabla \hat{J}^\varepsilon(\hat{u}^j + \alpha^j d^j)]^T d^j = 0$, which is

$$\alpha^j = -\frac{\sum_{k=0}^{K-1} u^j(k)^T d^j(k)}{\sum_{k=0}^{K-1} d^j(k)^T d^j(k)} - \frac{\sum_{k=0}^{K-1} \sum_{i=1}^{3} \hat{I}_i^j(k) d_i^j(k)}{2\varepsilon \sum_{k=0}^{K-1} d^j(k)^T d^j(k)} \tag{7.91}$$

Step 8) Update the control sequence, for the next iteration

$$u_i^{j+1}(k) = \begin{cases} 0 & u_i^j(k) + \alpha^j d_i^j(k) < 0 \\ u_{max} & u_i^j(k) + \alpha^j d_i^j(k) > u_{max} \\ u_i^j(k) + \alpha^j d_i^j(k), & elswhere \end{cases} \tag{7.92}$$

where $i = 1, 2, 3$ and $k = 0, 1, 2, \ldots K - 1$. Then, jump to Step 2 until the control sequence does not change any more.

It is worth mentioning that during the algorithm development, we notice that a poor discretization with small K and M leads to a poor approximation of the integrals for computing the Hamiltonian. Consequently, the algorithm does not converge as we expect. To improve its robustness, we use the following formula for integrals necessary for the Hamiltonian computation

$$\int_X \pi_i(x, k\Delta T) \rho_j(x, k\Delta T) dx \approx$$
$$\frac{\Delta X}{6} \sum_{m=1}^{M} \pi_i(m, k)(2\rho_j(m, k) + \rho_j(m + 1, k)) + \tag{7.93}$$
$$\frac{\Delta X}{6} \sum_{m=1}^{M} \pi_i(m + 1, k)(\rho_j(m, k) + 2\rho_j(m + 1, k))$$

This formula is derived using the following approximations for $x = m\Delta X$ and $x + \delta_x \in [m\Delta X, (m + 1)\Delta X]$

$$\rho(x + \delta_x, k\Delta T) \approx \rho_i(m) + \frac{\rho_i(m + 1) - \rho_i(m)}{\Delta x} \delta_x$$
$$\pi(x + \delta_x, k\Delta T) \approx \pi_i(m) + \frac{\pi_i(m + 1) - \pi_i(m)}{\Delta x} \delta_x$$

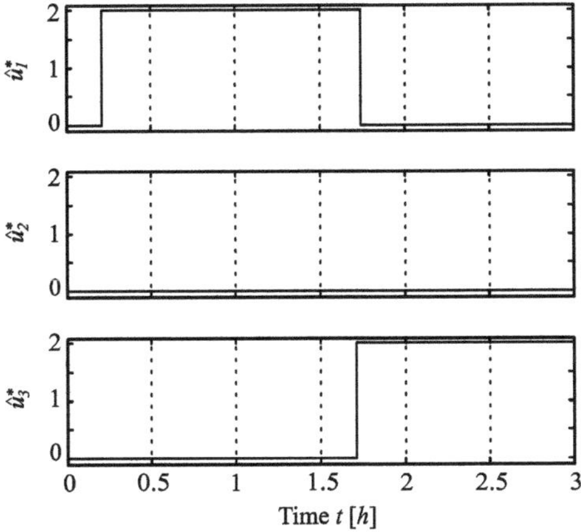

Fig. 7.7. Components of control $\hat{u}^* = [\hat{u}_1^* \ \hat{u}_2^* \ \hat{u}_3^*]^T$. This control is computed for the cost function $\hat{J}^\varepsilon$, $\varepsilon = 10^{-7}$, $\Delta T = 0.03h$, $u_i \in [0, 2]$, $i = 1, 2, 3$ [55], © 2006 IEEE.

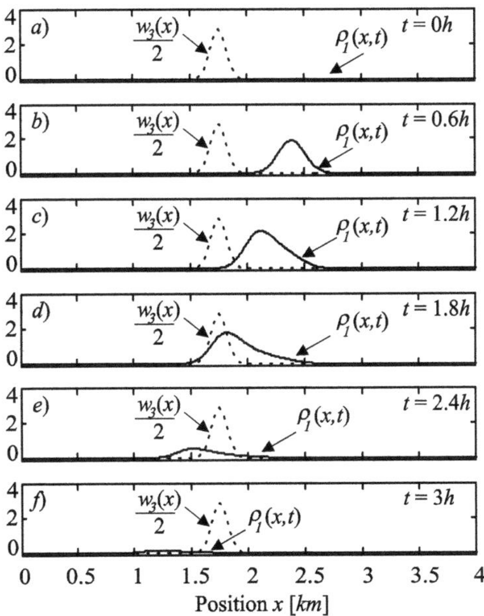

Fig. 7.8. State PDF evolution of the robots moving left $\rho_1(x, t)$ for $t = 0.6,\ 1,\ 1.2,\ 1.8,\ 2.4h$, $u_i \in [0, 2]$, $i = 1, 2, 3$ [55], © 2006 IEEE

We apply this more efficient gradient-based algorithm to our example under the same conditions, i.e., $\Delta T = 0.03$, $K = 100$, $M = 500$, the parameter $\varepsilon = 10^{-7}$ and the initial guess $\hat{u}^1(k) = [0.5\ 0.5\ 0.5] \ \forall k$. The convergence of the control sequence iterations towards the stationary point is improved. Now, the optimal control is reached after 22 iterations (see figure 7.6, solid line). This improvement

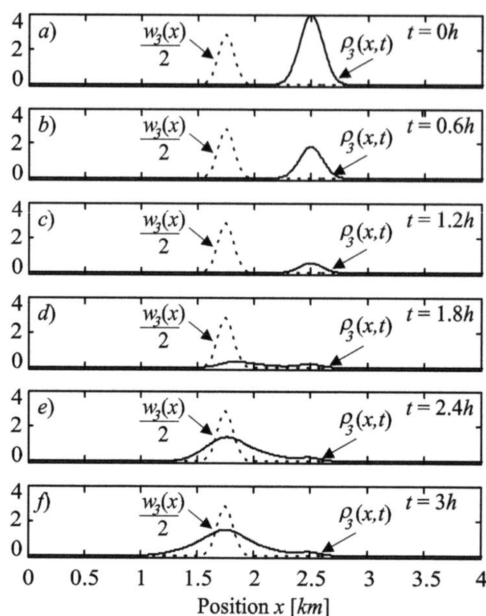

Fig. 7.9. State PDF evolution of stopped robots $\rho_3(x, t)$ for $t = 0.6,\ 1.2,\ 1.8,\ 2.4h$, $u_i \in [0, 2]$, $i = 1, 2, 3$ [55], © 2006 IEEE

is reached without involving the cost function, or its second order derivatives. The numerical complexity of this algorithm also scales with K.

The obtained control is identical to the one obtained for the simpler algorithm [55] shortly described in the previous subsection. Therefore, we will describe the control as in [55].

The components of control $\hat{u}^*$ are given in Fig. 7.7. The first component $\hat{u}_1^*$ starts with zero value, then it changes at $t = 0.21h$ to 2. Before it turns to 0 again, the control $\hat{u}_3^*$ has already changed from 0 to 2. Thus, between $t = 1.71h$ and $t = 1.74h$, $\hat{u}_1^*$ and $\hat{u}_3^*$ are both 2. This can be understood as an effective slowdown of the velocity of moving to the left, since each robot makes a transition to the *move left* and *stop* states with the highest possible rates. After $t = 1.74h$, $\hat{u}_1^*$ is 0 and $\hat{u}_3^*$ is 2 by the end of the time interval $[0, T]$. All the time, in this interval, $\hat{u}_2^*$ is zero, i.e., the optimal control does not include transitions to the discrete state *move right*. Since the initial PDF of robots moving right is $\rho_2(x, 0) = 0$, we can conclude that this PDF will be zero all the time, i.e, $\rho_2(x, t) = 0$, $\forall t \in [0, T]$. Therefore, only the evolution of $\rho_1(x, t)$ and $\rho_3(x, t)$ are presented in Fig. 7.8 and Fig. 7.9, respectively.

Starting with the initial PDFs $\rho_i(x, 0)$, $i = 1, 2, 3$, the control $\hat{u}^*$ produces distribution evolutions such that, at the terminal time $T = 3h$, the PDF $\rho_3(x, t = T)$, depicted in Fig. 7.9, has one peak value at the same position as $w_3(x)$ peak. The transition rates are limited, therefore, there are robots that have never moved from the discrete state *stop* at the terminal time T. This is shown in the plot of $\rho_3(x, T)$, where a small plateau exists for the x values corresponding to $\rho_3(x, 0)$ maximum.

We can see from Fig. 7.9 that our original intuition to produce a sharper robot distribution in the terminal time in comparison to the distribution in the initial time has failed. This is because the evolution of $\rho_3(x,t)$ is constrained by the population dynamics, i.e., the PDE system (7.18) and the corresponding initial condition (7.39).

The presented numerical example along with the derivation of this section shows that the Minimum Principle for partial differential equations can be exploited for the computation of the robotic population optimal control in the same way as Pontryagin Minimum Principle is used for the ODE optimal control computation [50], [78].

7.5 Summary

In this chapter we introduce the $CTMC\mu A$ of a large size robotic population. We present the capability of our modeling approach to predict the position PDF of the robotic population. This gives us motivation to formulate an optimal control problem in which the position PDF of the robotic population can be maximized over a region of robots operating space for a given time T. This optimal control problem is defined using the objective function based on the position PDF which must be minimized. We discuss the application of the Minimum Principle for PDEs to solve this optimal control problem.

Key points

- Modeling of the large size robotic population by a stochastic $CTMC\mu A$ model.
- Robotic population optimal control problem and the application of the Minimum Principle for partial differential equations to this problem.

8. Conclusions and Future Work

In this book we have explored common grounds for studying multi-agent populations of biological cells and robotic teams. The main motivation for this effort comes from the observation that both cells and robots are sophisticated agents with ability to sense the environment, analyze signals and act individually, or in cooperation with their teammates.

Our main concern in this monograph is the question on how dynamics of an individual agent propagates to the population dynamics. An answer to this question can provide an insight into behavior of individual agents from the measurements of the population properties. This opens an opportunity for a smart manipulation of the population properties. Therefore, the question we approached in this book is important for understanding biological data, as well as the manipulation, i.e., control of multi-agent systems, such as biological cell populations and large-size robotic teams.

The original contributions of this book can be shortly summarized in the following points:

- Development of the Micro-Agent model of individual agents and Stochastic Micro-Agent model of the agent population. This includes:
  - System of partial differential equations describing the state probability density function of a population which can be modeled by a *Continuous Time Markov Chain Micro-Agent*.
  - Modeling approach to the parameter uncertainty of the Micro-Agent continuous dynamics.
- Hybrid system approach to the modeling of TCR expression dynamics of an individual T-cell and T-cell populations. As part of this contribution, a novel data processing algorithm, which is able to reject the T-cell autofluorescence from the Flow Cytometry measurements, has been developed.
- Modeling a large-size robotic population, including the formulation of an optimal control problem taking the population to a desired location with the maximal probability; the algorithm for computing numerical optimal control for this problem.

In the rest of this chapter, we summarize our conclusions and discuss some possible future work related to the theoretical development of the *modeling framework*, the *modeling of biological cell populations* and the potential application to the *modeling and control of large-size robotic populations*.

D. Milutinovic and P. Lima: Cells & Robots, STAR 32, pp. 91–95, 2007.
springerlink.com

Modeling framework: In the approach to the modeling and analysis of a large population of agents we have presented here, each individual agent is described by a deterministic Hybrid Automaton model, named a *Micro-Agent*. This model describes an agent behavior which obeys ordinary differential equation dynamics in each discrete state. However, the discrete state can change as a result of an input event sequence and, independently, of the state of continuous dynamics. Application of this model to an individual robot or a cell describes them as deterministic devices. To model the complex environment surrounding agents we have assumed that their input event sequences are stochastic and we introduced a *Stochastic Micro-Agent* model of the population.

To answer the question about the relation between the individual agent dynamics and the macro-dynamics of the population observations, we found useful to apply statistical physics reasoning. Motivated by this reasoning, we described this relation by the agent state probability density function (PDF). The goal of our theoretical work was to infer the state PDF of the agent population and then, based on this PDF, to find the PDF of the population macro-observations. This resulted in a system of partial differential equations (PDE) describing the evolution of the state PDF when the Stochastic Micro-Agent population discrete state is modeled by a Markov Chain. Using the state PDF, we predicted the macro-observation PDF as the output PDF, in the case when it can be calculated as a sum of the state PDFs.

For the same case of Markov Chain discrete state transitions, we also introduced an approach to handle the problem of the parameter uncertainty of the Continuous Time Markov Chain $(CTMC)\mu A$. The approach aims towards a systematic procedure that transforms Hybrid Automata models without the uncertainty to models that include the uncertainty description. While we succeeded in the formulation of how to handle some types of the parameter uncertainty, which is described by a discrete PDF, we found it difficult to manage the parameter uncertainty, which is defined by a continuous PDF.

Extending the theory to the modeling of the agent population described by general Hybrid Automata, dealing with a more complex macro-observation model, or a proper treatment of the parameter uncertainty given by a continuous PDF are some possible directions for further theoretical development. We believe that, in this development, potentials of using tools for describing non-stationary thermodynamic systems have to be considered. A very interesting effort for further research would be the formulation of thermodynamic laws for the agent population described by Hybrid Automata.

Modeling a biological cell population: We applied the presented modeling approach to the membrane T-Cell Receptor (TCR) density dynamics of a T-cell population interacting with ligand-bearing APCs. Available analytical models are ordinary differential equation models of the TCR average amount of the population. These models assume that the population average amount dynamics is equivalent to the dynamics response of an average cell. Using only the average value of the measurements means that a lot of potentially useful information, being part of the experimental data, is discarded.

Our modeling approach incorporates all of this information into the model, in which each individual T-cell is described by a deterministic Micro-Agent. Under a stochastic assumption about the Micro-Agent input event sequence, the Stochastic Micro-Agent model of the T-cell population was introduced. The relationship between the proposed model and the time evolution of the state PDF provides us with the ability to predict the experimental membrane TCR amount distribution. This prediction is tested against the experimental data. The predicted distribution should not only match the mean value, but the overall shape of the experimentally obtained distributions. Therefore, we are capable of deciding what kind of functions should describe continuous dynamics of discrete states. In other words, we identify them within a function space.

Using the experimental data of interactions between T-Cells and antibodies under the conditions where only one discrete state exists, we tested linear and quadratic hypotheses of TCR expression decrease dynamics. We concluded that this dynamics should be described by the same linear function as the one we found in the previous T-Cell-APC experiment. Finally, we could identify the parameters of the dynamics.

In order to identify parameters, we relied on the Flow Cytometry, i.e., type of microscopy which provides the laser scan of each cell from the cell population sample. To measure the TCR amount level, the TCRs are labeled. After their exposition to the laser light, the cells emit the intensity of the light corresponding to the TCR amount. However, one part of this light is due to the autofluorescence. To obtain the shape of TCR amount distributions of the cell population, the autofluorescence component must be rejected. This algorithm is based on the QQ-plot for estimating distributions and the stable Richardson-Lucy deconvolution method.

During the development of the TCR expression model, we have made a few simplifications that are not free from controversy on biological grounds. First, we assumed that conjugate life-times and the waiting time for conjugation are both exponentially distributed. This assumption is reasonable for the waiting time, since T-cells and APCs seem to perform random walks, at least *in vitro*. As to the distribution of conjugate life-times, it seems to be akin to an exponential under some experimental settings [27], while in other conditions, it is more bell-shaped, with the mean of a few hours [31].

Another assumption, that is worth discussing, concerns the initial PDF of the T-cells that never conjugated. This initial membrane TCR density has already some variance, in contrast to the Dirac pulse one would expect according to the model of a T-cell that would always be free. This means that there is some stochastic process, which can be an interaction, as we modeled here, but that is, most likely, intrinsic to the internal machinery of the T-cell. In the latter scenario, the values of the rates of the conjugate formation and dissociation we obtained are most likely under-estimated. Note that one can compensate the presence of extra sources of variance by reducing the variance due to the stochasticity in the

APC conjugated formation and dissociation, which is achieved by increasing the rates of these processes.

Potential future theoretical work along these research lines includes an extension to T-cell population dynamics, where the life-history parameters, such as death and proliferation, are themselves functions of the TCR signaling. Also, initial and steady state distributions may be more carefully studied using better approximations or including better hypothesis about the TCR dynamics in the absence of APCs and extra sources of the variance in the membrane TCR density. On experimental grounds, it would be interesting to test the proposed model using experimentally obtained statistics of the stochastic event sequence.

The theory presented here can be used to provide additional insights into other biological phenomena, where cells or other biological agents alternate among discrete states.

Modeling and control of a robotic population: Here we have given a motivation for the application of our modeling approach to the modeling of large-size robotic populations. The proposed model of a large-size robotic population considers not only the probabilistic description of the task allocation, but also the distribution of the population over the operating space. The models of large-size populations used previously in the literature are related either to robotic formations, i.e., relative positions of the robots in the operating space, or to the task allocation and the task performance modeled under the probabilistic framework.

The motivating example we introduced illustrates that different model parameters lead to different densities of the robotic population in probabilistic space. Based on this, we formulated an optimal control problem in which an objective function includes the population state PDF. In this problem, the objective function reflects the mission goal. A possible choice for the objective function is the one which maximizes the robotic presence in a defined region. Since the time evolution of the state PDF obeys a PDE, this controller design is a PDE optimal control problem. To solve this problem, the Minimum Principle for PDEs [21], similar to Pontryagin Minimum Principle for ODE, can be applied.

The difficulty we faced in the application of the Minimum Principle for PDE originates from the problem of solving the PDE two-point-boundary-value problem and the singular condition for optimal control. Therefore, we also considered the possibility to solve the optimal control problem numerically. After discussing the complexity of possible solutions, we exploited results obtained from the application of the Minimum Principle to solve the problem numerically efficiently.

The result presented in this book opens the door to an opportunity to extend the research in several directions, including the single robot task composition, communication among the robots and extension of the modeling and control approach. It is worth mentioning that the distribution of the population evolves in the same way as the probability density function of a single robot. Therefore, the results presented here can be used for the modeling and control of one robot in probabilistic sense. This can be used for the trajectory planning and composition of robot task sequences in order to satisfy the mission objective.

In the case of a two-robot team, our model can explicitly include the rates of receiving information from the other robot. This can be exploited to estimate the necessary intensity of the communication between the two robots, while the two-robot team mission remains successful. Here we developed a full analytical model for the case when all the transitions over discrete states can be described by a Markov chain. The transition rates are equal for overall population. However, the transition rates can depend on the robot position, e.g., the intensity of communication errors depends on the relative position between the robot and the command source. In that sense, it would be of certain interest to extend our work to the case when the transition rates are fast-changing and unpredictable functions of the robot position.

In this work we only considered the centralized control of a robotic population and we designed the best controller having applied this strategy. For further improvement of the control performance, the local robot controllers have to be taken into account. Possibly, the best control strategy would be a centralized optimal control that provides coherent behavior of the population, and local robot controllers (decentralized control) to provide the performance improvement and robustness. As our future work, we are considering designing such a kind of controller under the framework introduced in this paper. We are also giving attention to extending the application of the PDE Minimum Principle to other types of optimal criteria, e.g., optimal control with infinite terminal time and different cost functions, which can control the shape of the state probability density function.

The final conclusion regarding the contents of this book is that we have made a step towards a mathematically tractable way of describing and possibly controlling complex systems composed of a large amount of agents. This step shows that cells and robot teams can be studied within the same theoretical framework. At this moment we can hardly say what the impact of the presented theory on the immune system or multi-agent (robotic) systems research will be. However, we are sure that our effort to exploit the cell-robot analogy can inspire future research in Biology and Multi-Agent systems. This can include understanding of spatio-temporal formations of cell populations, comprehending extra- and intracellular communication pathways, centralized versus decentralized decision making trade-off in Multi-Agent systems, design of energy efficient robot populations, as well as design of advanced medical treatments.

A. Stochastic Model and Data Processing of Flow Cytometry Measurements

Flow Cytometry is an observation method which is used to measure the TCR amount. This method is implemented in special devices, so-called Flow Cytometry scanners. There are prescribed cell treatments invented to increase the correlation between the scanner measurements and the observed quantitative property of the cells. The Flow Cytometry scanner working principle is depicted in Figure A.1.

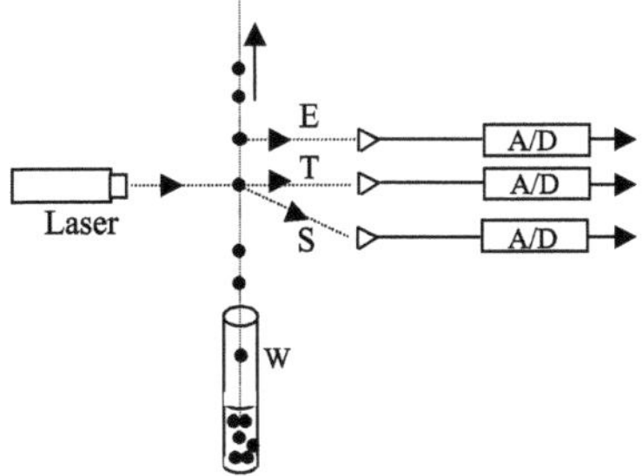

Fig. A.1. Flow Cytometry scanner working principle: W-T-cell container S-scattered light, T-transmitted light, E-emitted light, A/D-analog to digital converter

The vessel W contains the T-cells in suspension. The scanner has a special construction which allows it to sample exactly one cell and expose it to the laser light. For a different type of analysis a different wavelength can be applied. The intensity of the scattered light or the light emitted by the cell after the exposure (fluorescence) can be recorded in digital format after an analog to digital conversion. In our case the TCR molecules are labeled by fluorescent probe. Therefore, the fluorescence intensity has particular meaning because it corresponds to the TCR amount. Let us denote the intensity of the fluorescence intensity with z, then we can write:

$$z = \alpha x + \theta \tag{A.1}$$

D. Milutinovic and P. Lima: Cells & Robots, STAR 32, pp. 97–105, 2007.
springerlink.com   © Springer-Verlag Berlin Heidelberg 2007

where x is the amount of TCRs, α is a constant parameter and θ is the T-cell autofluorescence component of the measured signal. The Flow Cytometry result is a distribution N_m of the signal z measurements over the T-cell population. An example of real data N_m is given in Figure A.2. In this figure, the values of the signal z are amplified by a logarithmic amplifier sampled by 10bits A/D converter. The horizontal axis in this figure is a log_{10}-scale intensity of the fluorescence light and the vertical axis is the quantity of cells given the intensity of the light. Using the observation of the T-cell without the TCRs with the fluorescent probe, the T-cell autofluorescence distribution N_θ can be measured. An example of this measurement is presented in Figure A.2.

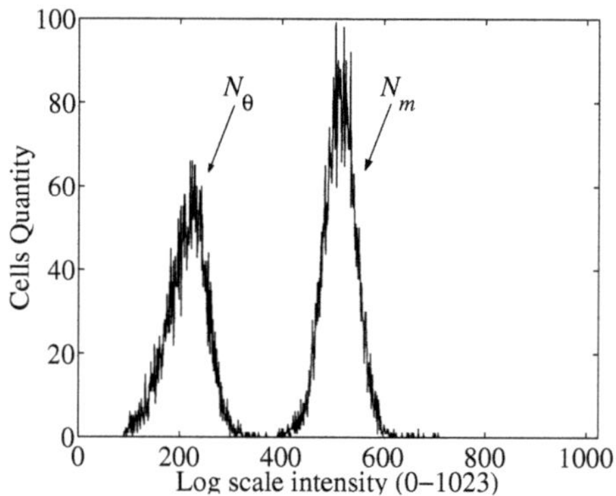

Fig. A.2. Flow Cytometry measurements: N_m and N_θ, by Lino [45]; the full scale from 0–1023 covers 4 decades

Taking the distributions measured by Flow Cytometry, N_m and N_θ, we can estimate the PDFs of the measured signal z (p_m) and autofluorescence signal θ (p_θ). There are different algorithms which can be used for that estimation and, in the sequel, we present one of the methods we find suitable. However, to test our $CTMC\mu A$ model, we should compare our prediction of TCR PDF to TCR PDF which is estimated from experimental data. Therefore, we have the following problem: *How to estimate the unknown TCR PDF p_x given p_m and p_θ?*

The starting point to solve this problem is equation (A.1). Assuming that $\alpha = 1$, then:

$$z = x + \theta \tag{A.2}$$

from which we conclude that:

$$p_m(z) = \int_{-\infty}^{+\infty} p_\theta(z - x)p_x(x)dx = p_\theta(z) * p_x(z) \tag{A.3}$$

The PDF of the measured signal is the result of the convolution (symbol *) between autofluorescence and TCRs PDF. To calculate $p_x(z)$ we should provide an algorithm which causes *deconvolution* of the noise PDF (p_θ) from the measurement signal PDF (p_m). Deconvolution is an operation and its solution can be non-unique. Because of that the numerical algorithm which solves the deconvolution can be very sensitive to small difference in input data. In this section, we will present an algorithm for the stable deconvolution, called the Richardson-Lucy deconvolution algorithm [47, 66].

The complete algorithm to estimate TCRs PDF p_x based on the N_m and N_θ experimental data is depicted in Figure A.3. Using the rough experimental data the PDF p_m and p_θ should be estimated, then the deconvolution algorithm, which takes out p_θ from p_m, produces the TCRs PDF p_x. In the rest of this section, the PDF estimation and deconvolution algorithm we are using will be presented.

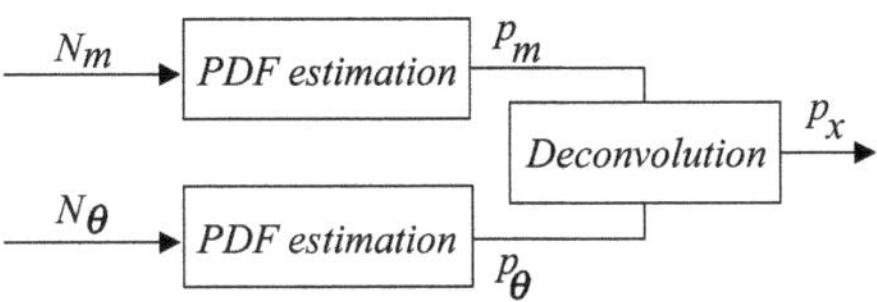

Fig. A.3. The Flow Cytometry data processing measurements: N_m - fluorescence distribution, N_θ - autofluorescence distribution, p_m - estimated PDF of fluorescence, p_θ - estimated autofluorescence PDF

A.1 Probability Density Estimation Algorithm

The probability density estimation algorithm used in this work is based on the so-called QQ-plot [18, 19]. This algorithm estimates the PDF as a weighted sum of the kernel probability density functions:

$$\hat{p} = \sum_{k=1}^{l} c_k \varphi(m_k, \sigma_k) \tag{A.4}$$

where c_k is a weighting coefficient and $\varphi(m_k, \sigma_k)$ is a PDF with a given mean value m_k and variance σ_k. Although the PDF parameters are its mean value and variance, this algorithm is not constrained to the case where φ is Gaussian. The type of kernel functions can be chosen appropriately, while the number of kernel functions is computed using an optimization procedure. Since this algorithm computes a number of kernel functions automatically, we find this algorithm suitable for our application. In the following text we will describe the algorithm to compute unknown parameters c_k, m_k and σ_k given data samples and PDF φ.

Let us assume that the data samples are given by the sequence of measurements $\{x_i\}$, $i = 1, 2, \ldots I$. By sorting this sequence of real numbers in nondecreasing order, we can produce an ordered sample sequence $\{y_i\}$, which satisfies $i < j \Rightarrow y_i \leq y_j \ \forall i, j \in \{1, 2, \ldots I\}$. To generate QQ-plot, the data samples PDF $p(\cdot)$, i.e., corresponding cumulative distribution P, have to be assumed. The QQ-plot is the plot of ordered sequence $\{y_i\}$ versus $P^{-1}(r_i)$, where:

$$r_i = \frac{i-1}{I-1} \tag{A.5}$$

The probability that some observation ν will have index i, in the ordered sample sequence $\{y_i\}$ is:

$$Pr(i|\nu) = \binom{I-1}{i-1} P^{i-1}(\nu)(1 - P(\nu))^{i-i} \tag{A.6}$$

and the expected value of index i, given observation ν, is:

$$m_{i|\nu} = E\{i|\nu\} = 1 + (I-1)P(\nu) \tag{A.7}$$

If $\nu = y_i$, then

$$i \approx m_{i|y_i} = 1 + (I-1)P(y_i) \Rightarrow y_i = P^{-1}\left(\frac{i-1}{I-1}\right) \tag{A.8}$$

Equation (A.8) shows why the QQ-plot is originally introduced as a test of data samples cumulative distribution hypothesis P. If the hypothesis P is correct, then the QQ-plot is a straight line. In the case of the zero mean hypothesis, with unit variance P_0 for the random variable $\frac{y_i - m}{\sigma}$, where $m = E\{y_i\}$ and $\sigma = E\{(y_i - m)^2\}$, the equation (A.8) is:

$$\frac{y_i - m}{\sigma} = P_0^{-1}(r_i) \tag{A.9}$$

i.e.,

$$y_i = m + \sigma P_0^{-1}(r_i) \tag{A.10}$$

Given an ordered data sample sequence and P_0 hypothesis, Equation (A.10) can be used to estimate the PDF using the least squares (LS) algorithm. If the QQ-plot is linear, the estimated value m and σ together with P_0 define the estimated PDF. In the non-linear case, the algorithm [18, 19] makes the piecewise linear approximation of the QQ-plot and applies LS estimation (A.10) to each plot segment. The result is a sequence of m_k and σ_k corresponding to linear segments sequence $a^* = a_k^*$. The first and the last point of each linear segment are $(P_0(r_{s_k^*}), y_{s_k^*})$ and $(P_0(r_{e_k^*}), y_{e_k^*})$, where s_k^* and e_k^* are the first and the last index of the data corresponding to linear segment a_k^*. If P_0 is $\varphi(0, 1)$, the estimated PDF is

$$\hat{p} = \sum_{k=1}^{l} c_k \varphi(m_k, \sigma_k) \tag{A.11}$$

where

$$c_k = \frac{e_k^* - s_k^*}{I - 1} = \frac{n_k^*}{I - 1} \tag{A.12}$$

so that the weighting factor c_k is the ratio between the number of points inside the linear segment a_k^* ($n_k^* = e_k^* - s_k^*$) and I which is the total number of points composing the QQ-plot.

Obviously, the piece-wise linear approximation is not unique. The complete solution of the piece-wise approximation is to decide the number of segments, as well as the first and the last point of each segments. The proposed suboptimal algorithm [18, 19] computes the optimal approximation a^*, which minimizes the criterion very similar to the minimum description length criterion (MDL):

$$J = \sum_{k=1}^{n_a} d(a_k, QQ) + \gamma n_a \log I \qquad (A.13)$$

where

$$d(a_k, QQ) = \frac{1}{n_k} \sum_{p=1}^{n_k} \left(a_k^j(P_0^{-1}(r_p)) - QQ(P_0^{-1}(r_p)) \right)^2 \qquad (A.14)$$

$d(a_k, QQ)$ is a distance between linear segment a_k and QQ-plot, n_k denotes the number of QQ-plot points inside the k-th linear segment and γ is a positive weighting parameter to the number of linear segments . If the parameter $\gamma = 0$, the criterion minimizes the distance between the piece-wise linear approximation and the QQ-plot, which results in a large number of the linear segments containing few data points. The piece-wise linear approximation is very good, but the mean value and variance of the linear segments are not statistically justified because of the few data points segment. Giving weight to the second term of criterion, $n_a \log I$, the trade-off between the number of linear segments and the number of points per segment can be made. An extensive numerical simulation suggests $\gamma \in (0, 0.05)$ [18, 19]. The complete QQ-plot based PDF estimation algorithm can be described as follows[18, 19]:

step 1) in the initial step ($j = 1$), the QQ-plot is approximated by only one linear $a_1^1 = (1, I)$ segment determined by the first and the last point of the QQ-plot.

step 2) among all the segments in the segmentation find the segment which is the worst approximation of the QQ-plot. The rule is to calculate the distance $d(a_k^j, QQ)$ to each segment, where a_k^j is the kth linear segment of the jth iteration. Denote by M the index of the segment corresponding to the maximal value of distance $d(a_k^j, QQ)$:

$$d(a_k^j, QQ) \leq d(a_M^j, QQ), \forall k, k, M \in 1..n_a^j \qquad (A.15)$$

where n_a^j is the number of linear segments in the iteration j. Now, divide the linear segment a_M^j into two segments. The point which divides the new segments is given by the index:

$$i_d = int \left(\frac{s_M^j + e_M^j}{2} \right) \qquad (A.16)$$

where s_M^j and e_M^j are the indexes of the first and the last point of the segment a_M^j and int is the function which calculates the integer value of a real number. The new linear segment sequence a^{j+1} have $n_a^{j+1} = n_a^j + 1$ linear segments.

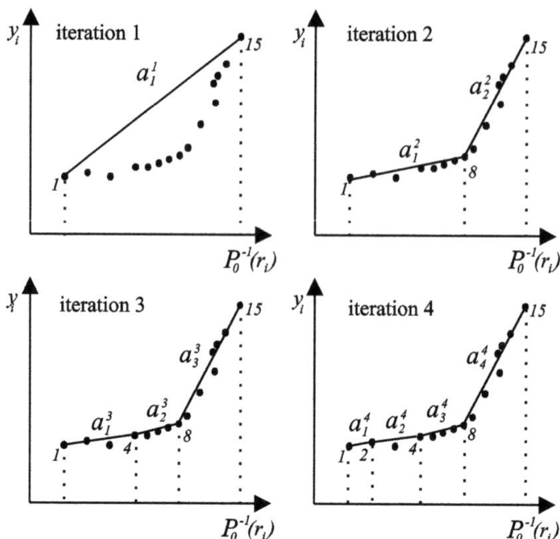

Fig. A.4. Suboptimal algorithm for the piece-wise linear approximation of the QQ-plot [18]

step 3) compute the criterion of the jth J^j and of $j+1$-th iteration J^{j+1}, if the $J^j > J^{j+1}$, go to *step 2)* and increase counter j by 1; otherwise, finish the algorithm and stop procedure with $a^* = a^j$.

Figure A.4 illustrates this algorithm. The QQ-plot is composed of 15 data points. In the initial iteration, the QQ-plot is approximated by only one linear segment. This linear segment is divided into two segments with equal number of points. In the second iteration, the distance of the linear segments to the QQ-plot (A.14) is calculated for each segment. The segment a_1^2, with the larger distance, is split in two for the next iteration. The same is applied in the third iteration, and so forth.

An example of the PDF estimation applied to Flow Cytometry measurements is presented in Figure A.5. As it is already explained in this section, the N_m and N_θ are measured using the logarithm $\log_{10}$ of intensity. Therefore the QQ estimates are denoted by $p_m^{\log}$ and $p_\theta^{\log}$. The parameters c_k, m_k and σ_k determined for the distributions are given in Tables B.2 and B.1(Appendix B), respectively. However, we should notice that the deconvolution algorithm is applicable only to the PDFs defined over a linear scale. The transformation which converts the PDF in log_{10} scale p^{log} to the PDF in linear scale p is:

$$p(z) = \frac{1}{z \ln 10} p^{log}(log z) \qquad (A.17)$$

The first and the last points to which we apply this transformation are $log z = 0$ and $log z = 1023$ corresponding to $z = 1$ and $z = 10^4$, respectively. The estimated PDF p_m and p_θ corresponding $p_m^{\log}$ and $p_\theta^{\log}$ (A.5) are presented in Figure A.6.

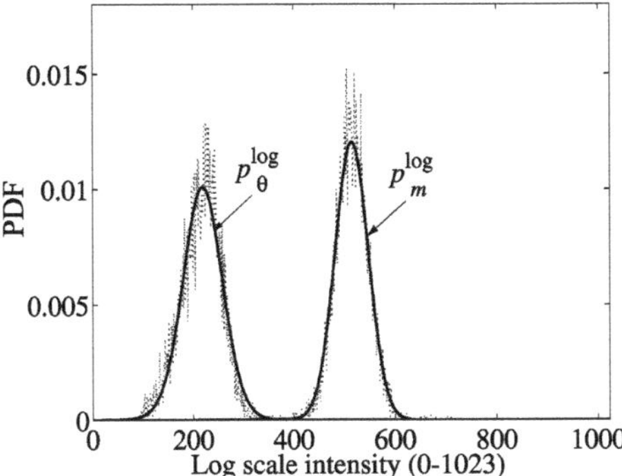

Fig. A.5. Example of the PDF estimation: p_m^{log} - PDF which corresponds to non-stimulated T-cells, p_θ^{log} - PDF of autofluorescence; the dotted lines in the background are the normalized N_m and N_θ measurements

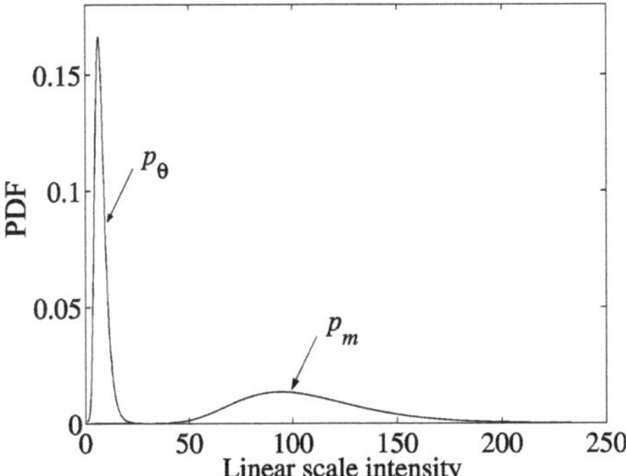

Fig. A.6. Example of the PDF estimation: p_m - PDF which corresponds to non-stimulated T-cells, p_θ - PDF of autofluorescence

A.2 Richardson-Lucy Deconvolution Algorithm

There are different deconvolution methods that can be used to compute the unknown distribution p_x of TCR receptors using the PDF of measurements p_m and the autofluorescence PDF p_θ. Most of the methods are developed for the restoration of the astronomical images degraded due to the optical effects. This problem and its possible solution are presented in [66], which introduces the Richardson-Lucy (RL) iterative scheme used in our work. In [47], some properties of this algorithm are derived. We choose to apply this algorithm as it is numerically stable [28].

The RL method is based on Bayes rule. Writing the TCRs PDF as follows:

$$p_x(x) = \int_{-\infty}^{\infty} p(x|z)p_m(z)dz \qquad (A.18)$$

and applying Bayes rule:

$$p(x|z) = \frac{p(z|x)p_x(x)}{\int_{-\infty}^{\infty} p(z|x)p(x)dx} \tag{A.19}$$

and substituting in equation (A.18), we have

$$p_x(x) = \int_{\infty}^{\infty} \frac{p(z|x)p_x(x)}{\int p(z|x)p_x(x)} p_m(z)dz \tag{A.20}$$

On the right side of equation (A.20), $p_x(x)$ is also the solution we need. In many applications of Bayes theorem, when the term $p_x(x)$ is not initially known, an accepted practice is to make the best of a bad situation and use an initial estimate of the $p_x(x)$ [66]. When this practice is applied to the equation (A.20), we obtain the iterative formula:

$$p_x^{r+1}(x) = p_x^r(x) \int_{-\infty}^{\infty} \frac{p(z|x)}{\int p(z|x)p_x^j(x)} p_m(z), r = 0, 1, 2, \ldots \tag{A.21}$$

where $p_x^r(x)$ is the PDF of the rth iteration. To start the iteration, the initial PDF $p_x^0(x)$ should be estimated. In the absence of any information about $p_x^0(x)$, the uniform distribution can be taken. Using the PDF data samples at x_i and z_k, the iterative algorithm is:

$$p_x^{r+1}(x_i) = p_x^r(x_i) \sum_k \frac{p(z_k|x_i)}{\sum_j p(z_k|x_j)p_x^r(x_j)} p_m(z_k), \tag{A.22}$$

$$r = 0, 1, 2, \ldots, \quad i = \{1, I\}, \quad j = \{1, J\}, \quad k = \{1, K\}$$

Considering the finite data samples of the PDF $p(z_k|x_i) = p_\theta(z_k - x_i) = S_{k-i+1}$, we can write:

$$p_x^{r+1}(x_i) = p_x^r(x_i) \sum_{k=i}^{c} \frac{S_{k-i+1}p_m(z_k)}{\sum_{j=a}^{b} S_{k-j+1}p_x^r(x_j)} \tag{A.23}$$

where $a = max(1, k - J + 1)$, $b = min(k, I)$, $c = i + J - 1$. When the uniform distribution is used for the initial estimate, equation (A.23) becomes:

$$p_x^1(x_i) = \sum_{k=i}^{c} \frac{S_{k-i+1}p_m(z_k)}{\sum_{j=a}^{b} S_{k-j+1}} \tag{A.24}$$

In our application, p_m is a smooth function which is the result of the QQ-plot-based estimation. The result of rth iteration is also a smooth function and since the iteration (A.23) converges [47] to obtain the result of deconvolution, the iteration (A.23) has to be repeated till some criterion for stopping the iteration is satisfied. In our work, we are using 800 data samples (I=K=800) and 10 iterations to compute a result. The maximum of difference between the results after the 10th and 11th iteration is 10^{-3}, which can be neglected. In the above discussion, we assume that $\alpha = 1$. If this is not the case, then p_x is the distribution of αx, where x is the amount of TCRs and α is a constant parameter. The p_x computed by the RL deconvolution of p_θ from p_m (Figure A.6) is presented in Figure A.7.

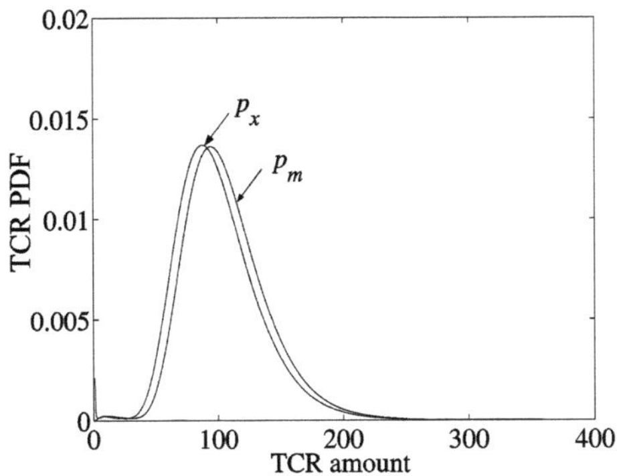

Fig. A.7. Example of the PDF deconvolution: p_x is received by the deconvolution of p_θ from p_m, Figure A.6: p_m- PDF which corresponds to non-stimulated T-cells, p_x-TCR PDF ($\alpha = 1$)

B. Estimated T-Cell Receptor Probability Density Function

In this appendix, the fluorescence intensity PDF estimations received from the experiment, where the T-cell population is exposed to $0.1\mu g$ of antibodies, are listed [45]. The experimental PDF of TCRs in Chapter 5.4, (Figure 5.11) are estimated using the data presented in the tables below and the algorithm in appendix A.

The PDF which corresponds to T-cell autofluorescence PDF p_θ^{log} and TCR distribution PDF p_m^{log} are estimated on a scale 0-1023 which covers 4 decades of the fluorescence intensity. The PDF estimation is received using the QQ-plot algorithm and the following form of PDF:

$$\hat{p} = \sum_{k=1}^{l} c_k N(m_k, \sigma_k) \tag{B.1}$$

where $\hat{p}$ is the PDF estimation, c_k is the weighting coefficient and $N(m_k, \sigma_k)$ is Gaussian PDF with mean value m_k, variance σ_k.

Data on TCR triggering and down-regulation mediated by anti-CD3 have been provided by Andreia Lino and Jorge Carneiro (for experimental details see [45]).

Table B.1. Estimated T-cell autofluorescence PDF, p_θ^{log}

k	c_k	m_k	σ_k
1	0.500000	219.006973	44.737006
2	0.250000	218.719874	35.587063
3	0.125000	219.067239	34.202273
4	0.062500	219.924756	33.367209
5	0.031153	221.148242	32.757356
6	0.015576	206.667384	40.546449
7	0.007788	214.446946	36.930977
8	0.007983	130.596407	70.848579

D. Milutinovic and P. Lima: Cells & Robots, STAR 32, pp. 107–110, 2007.
springerlink.com

Table B.2. Estimated initial TCR PDF, p_m^{log}

k	c_k	m_k	σ_k
1	0.499923	516.100516	33.092744
2	0.249962	515.060709	31.559593
3	0.124981	513.274641	33.801055
4	0.062490	512.669595	34.009551
5	0.031245	511.963582	34.266822
6	0.015700	509.312369	35.669370
7	0.007850	519.009316	31.150042
8	0.007850	345.113316	100.441415

Table B.3. Estimated TCR PDF p_m^{log} after 1min

k	c_k	m_k	σ_k
1	0.500000	508.674885	32.344656
2	0.250000	508.235393	33.070512
3	0.125000	508.423846	33.333285
4	0.062500	503.789116	37.341254
5	0.031088	509.100424	33.824941
6	0.015544	496.492949	40.325155
7	0.015868	398.519407	83.520207

Table B.4. Estimated TCR PDF p_m^{log} after 15min

k	c_k	m_k	σ_k
1	0.499811	476.475513	28.399256
2	0.250094	476.402118	27.760191
3	0.125047	474.160060	30.816296
4	0.062335	472.857118	32.543326
5	0.031356	481.272582	27.090544
6	0.015489	449.662944	44.245995
7	0.007934	477.490225	31.282866
8	0.003778	420.698779	54.453897
9	0.004156	242.619046	120.385419

Table B.5. Estimated TCR PDF p_m^{log} after 30min

k	c_k	m_k	σ_k
1	0.500000	438.626824	35.477503
2	0.249827	434.936595	28.304885
3	0.125087	435.563738	27.357542
4	0.062370	435.810992	27.617439
5	0.031185	414.737111	41.013147
6	0.015593	412.453985	43.053564
7	0.007970	348.147525	72.945346
8	0.007970	234.637843	117.938257

Table B.6. Estimated TCR PDF p_m^{log} after 45min

k	c_k	m_k	σ_k
1	0.499821	429.523135	38.074392
2	0.250089	426.046057	26.932232
3	0.125045	424.542408	28.127297
4	0.062522	423.186385	29.312424
5	0.031083	417.914223	33.319361
6	0.015720	410.279816	37.692587
7	0.007860	376.560978	53.633988
8	0.007860	148.905367	147.066739

Table B.7. Estimated TCR PDF p_m^{log} after 59min

k	c_k	m_k	σ_k
1	0.499826	406.815745	36.499304
2	0.249913	404.534681	28.818082
3	0.125130	404.078543	29.496719
4	0.062565	401.795267	31.496399
5	0.031283	409.086035	26.897037
6	0.015641	395.824537	34.568911
7	0.007647	343.507064	58.044419
8	0.003823	243.562673	100.770476
9	0.004171	5.746770	195.479308

Table B.8. Estimated TCR PDF p_m^{log} after 93min

k	c_k	m_k	σ_k
1	0.500000	346.909437	47.118715
2	0.249847	342.256061	35.006830
3	0.124923	340.608304	37.285647
4	0.062462	344.673575	34.043459
5	0.031231	320.589365	49.688884
6	0.015615	305.059566	57.427922
7	0.015922	62.725415	166.291622

Table B.9. Estimated TCR PDF p_m^{log} after 122min

k	c_k	m_k	σ_k
1	0.062341	380.768122	66.164479
2	0.062659	347.193873	45.427253
3	0.125000	338.657707	37.990182
4	0.250000	339.129183	38.576072
5	0.250000	339.083687	36.432553
6	0.125000	339.533806	35.380050
7	0.062341	335.337432	38.774265
8	0.031170	333.030683	39.699283
9	0.015585	336.137351	38.725043
10	0.007952	330.261145	41.846466
11	0.007952	142.109024	119.229166

C. Steady State T-Cell Receptor Probability Density Function and Average Amount

In this appendix, the equations which are used in the TCR PDF prediction of Sections 5.3 and 6.3 are presented. First, we show the set of equations describing the output PDF and the state PDF evolution of the original $CTMC\mu A$ model, then we show the same type of results for the $\overline{CTMC\mu A}$ model, which includes the parameter k_2 discrete parameter uncertainty.

In the case of the $CTMC\mu A$ without uncertain parameter k_2 (see Figure 6.4), the TCR distribution is calculated as:

$$\eta(x,t) = \sum_{i=1}^{3} \rho_i(x,t) \tag{C.1}$$

where the state PDF $\rho(x,t) = [\rho_1(x,t)\ \rho_2(x,t)\ \rho_3(x,t)]^T$ satisfies:

$$\frac{\partial \rho_1}{\partial t} = -\lambda_{12}\rho_1 - \nabla \cdot (f_1\rho_1) \tag{C.2}$$

$$\frac{\partial \rho_2}{\partial t} = \lambda_{12}\rho_1 - \lambda_{23}\rho_2 + \lambda_{32}\rho_3 - \nabla \cdot (f_2\rho_2) \tag{C.3}$$

$$\frac{\partial \rho_3}{\partial t} = \lambda_{23}\rho_2 - \lambda_{32}\rho_3 - \nabla \cdot (f_3\rho_3) \tag{C.4}$$

and the initial condition

$$\rho(x,0) = [1\ 0\ 0]^T \frac{1}{2\sigma_0 x \sqrt{(\pi)}} e^{-\frac{(M_0 - \ln(x))^2}{2\sigma_0^2}} \tag{C.5}$$

where

$$M_0 = \ln(100), \ \sigma_0 = 0.19 \tag{C.6}$$

In the case of $\overline{CTMC\mu A}$ model where k_2 parameter uncertainty is included (see Figure 6.5), the TCR distribution is calculated as:

$$\eta(x,t) = \rho_1(x,t) + \rho_3(x,t) + \sum_{i=1}^{3} \rho_{i\{2,i\}}(x,t) \tag{C.7}$$

D. Milutinovic and P. Lima: Cells & Robots, STAR 32, pp. 111–112, 2007.
springerlink.com

where evolution of the state PDF $\rho(x,t) = [\rho_1(x,t) \ \rho_2(x,t) \ \rho_3(x,t) \ \rho_4(x,t) \ \rho_5(x,t)]^T$ satisfies:

$$\frac{\partial \rho_1}{\partial t} = -(\lambda_{1\{2,1\}} + \lambda_{1\{2,2\}} + \lambda_{1\{2,3\}})\rho_1 - \nabla \cdot (f_1\rho_1) \tag{C.8}$$

$$\frac{\partial \rho_{\{2,1\}}}{\partial t} = \lambda_{1\{2,1\}}\rho_1 - \lambda_{\{2,1\}3}\rho_{\{2,1\}} + \lambda_{3\{2,1\}}\rho_3 - \nabla \cdot (f_2^1\rho_{\{2,1\}}) \tag{C.9}$$

$$\frac{\partial \rho_{\{2,2\}}}{\partial t} = \lambda_{1\{2,2\}}\rho_1 - \lambda_{\{2,2\}3}\rho_{\{2,2\}} + \lambda_{3\{2,2\}}\rho_3 - \nabla \cdot (f_2^2\rho_{\{2,2\}}) \tag{C.10}$$

$$\frac{\partial \rho_{\{2,3\}}}{\partial t} = \lambda_{1\{2,3\}}\rho_1 - \lambda_{\{2,3\}3}\rho_{\{2,3\}} + \lambda_{32,3\}}\rho_3 - \nabla \cdot (f_2^3\rho_{\{2,3\}}) \tag{C.11}$$

$$\frac{\partial \rho_3}{\partial t} = \lambda_{\{2,1\}3}\rho_{\{2,1\}} + \lambda_{\{2,2\}3}\rho_{\{2,2\}} + \lambda_{\{2,3\}3}\rho_{\{2,3\}} \tag{C.12}$$
$$-(\lambda_{3\{2,1\}} + \lambda_{3\{2,2\}} + \lambda_{3\{2,3\}})\rho_3 - \nabla \cdot (f_3\rho_3)$$

and the initial condition : $\rho(\mathbf{x},0) = \begin{bmatrix} 1 & 0 & 0 & 0 & 0 \end{bmatrix}^T \dfrac{1}{2\sigma_0 \mathbf{x}\sqrt{(\pi)}} e^{-\frac{(M_0 - \ln(\mathbf{x}))^2}{2\sigma_0^2}}$

$$\tag{C.13}$$

The boundary condition in both cases is

$$\rho(0,t) = \rho(10^3,t) = 0, \forall t \tag{C.14}$$

and the prediction of the steady state $\eta^s(x)$ is computed by brute force, leaving that t in the computations takes a large value. The average TCR amount $\bar{\eta}(t)$ is calculated using

$$\bar{\eta}(t) = \int_{-\infty}^{\infty} x\eta(x,t)dx \tag{C.15}$$

where $\eta(x,t)$ is computed for all t by (C.1) and (C.7) taking into account corresponding state PDF evolution for $CTMC\mu A$ (C.2)–(C.4) and $\overline{CTMC\mu A}$ (C.8)–(C.12), respectively.

D. Optimal Control of Partial Differential Equations

Since the state PDF evolution of the robotic population $CTMC\mu A$ model is described by a system of PDEs, the optimal control problem introduced in Section 7 should be considered as an optimal control problem of PDEs. The main result regarding this optimal control problem is the so-called *Minimum Principle* for PDEs.

In our case, we consider a more specific control problem PDE system which is in the form of the *Evolution equation*:

$$\rho'(t) = A(t)\rho(t) + f(\rho(t), u(t), t), \quad \rho(0) : \text{given} \tag{D.1}$$

and an objective function to be maximized

$$\rho_0(u, T) = g_0(\rho, \rho(u, T)) + \int_0^T f_0(\rho(u, \tau), u(\tau), \tau)d\tau \tag{D.2}$$

The PDE, as well as the objective function are written using the notation employed in the control theory of PDE [21], namely the *abstract form*, where $\rho'(t)$ is the time derivative of $\rho(t)$ and $\rho(t)$ is not only a function of time but also a function of the space variable. With some exceptions, where we use the reference for PDE [20], all the results presented in this appendix are from [21].

First, we will introduce the definitions and the hypothesis which are necessary to define the properties of the *Evolution equation* for which the *Minimum Principle* can be applied.

Definition D.1 [20]. The $L^p(X)$ space of functions is

$$L^p(X) = \{\rho : X \to R| \text{ is Lebesgue meas.}, \ \|\rho\|_{L^p} < \infty\} \tag{D.3}$$

where

$$\|\rho\|_{L^p} = \left(\int_X |\rho|^2 dx\right)^{\frac{1}{2}}, \ 1 < p < \infty \tag{D.4}$$

Definition D.2 [21]. *Tangent cone* $T_{Y_\rho}(\bar\rho)$ to Y_ρ at $\bar\rho$ consists of all ω such that, for every sequence $\{h_k\} \subset R_+$ with $h_k \to 0$ and for every sequence $\{\bar\rho_k\} \subset Y_\rho$ with $\bar\rho_k \to \bar\rho$, there exists a sequence $\{\rho_k\} \subset Y_\rho$ with $\rho_k \to \bar\rho$ and

D. Milutinovic and P. Lima: Cells & Robots, STAR 32, pp. 113–116, 2007.
springerlink.com

$$\frac{\rho_k - \bar{\rho}_k}{h_k} \to \omega \text{ as } k \to \infty \tag{D.5}$$

Definition D.3 [21]. *Negative polar cone* of subset $Z \in E$ is the subset Z^- of the dual space E^T defined by

$$Z^- = \{\rho^T \in E^T; \langle \rho^T, \rho \rangle \leq 0, \ \rho \in Z\} \tag{D.6}$$

where symbol $\langle \cdot, \cdot \rangle$ is the duality pairing of E and E^T.

Definition D.4 [21]. *Normal cone* $N_{Y_\rho} \subset E^*$ to Y_ρ at $\bar{\rho}$ is the (closed,convex) cone defined by

$$N_{Y_\rho}(\bar{\rho}) = T_{Y_\rho}(\bar{\rho})^- \tag{D.7}$$

Definition D.5 [21]. *Normal cone* $N_{Y_\rho} \subset E^*$ to Y_ρ at $\bar{\rho}$ is the (closed,convex) cone defined by

$$N_{Y_\rho}(\bar{\rho}) = T_{Y_\rho}(\bar{\rho})^- \tag{D.8}$$

Definition D.6 [21]. (*Patch complete*) The set of admissible control in the time interval $[0, T]$ taking values from the set U, $C_{ad}(0, T; U)$ is *patch complete* if every patch perturbation of the control $u \in C_{ad}(0, T; U)$ with arbitrary m and $v = (v_1, v_2, \ldots, v_m)$, $v_k \in C_{ad}(0, T; U)$ belongs to $C_{ad}(0, T; U)$.

The *patch perturbation* of the control u corresponding to e and v is:

$$u_{e,v}(t) = \begin{cases} v_j(t), \ t \in e_j, j = 1, 2, \ldots, m \\ u(t) \qquad elsewhere \end{cases} \tag{D.9}$$

Definition D.7 [21]. (*Saturated space*) The sequence of control $u_n \subset C_{ad}(0, T; U)$ is *stationary* if there exists a set e with $|e| = 0$ such that for every $t \in [0, T] \backslash e$ there exists n (depending on t) such that $u_n(t) = u_{n+1}(t) = u_{n+2}(t) = \ldots$. The space $C_{ad}(0, T; U)$ is called *saturated* if the limit of every stationary sequence belongs to $C_{ad}(0, T; U)$.

Definition D.8 [21]. (*Spike complete*) The set of admissible control in the time interval $[0, T]$ taking values from set U, $C_{ad}(0, T; U)$ is *spike complete* if every spike perturbation of the control $u \in C_{ad}(0, T; U)$.

The *spike perturbation* of the control u corresponding to s, h and v is:

$$u_{s,h,v}(t) = \begin{cases} v, \ s - h < t \leq s \\ u(t) \ \ elsewhere \end{cases} \tag{D.10}$$

where $0 < s \leq T$, $0 \leq h \leq s$ and v is element of U.

Definition D.9 [20]. Set E is separable if E contains a countable dense subset.

Definition D.10 [21]. (*Regular set*) Let $\phi(t,v)$ be a Banach space valued function defined in $[0,T] \times U$ such that $\phi(t,v)$ is integrable for v fixed. The *left Lebesgue set* of $\phi(t,U)$ is the set d of all $t \in [0,T]$ such that t is a left Lebesgue point of $\phi(t,v)$ for every $v \in U$. The $\phi(t,v)$ is *regular* in U if its left Lebesgue set has full measure in $[0,T]$.

Hypothesis under which Theorem D.1 is valid : The Fréchet derivative $\partial_\rho f_0(\rho,u,t)$ and $\partial_\rho f(\rho,u,t)$ exist in $[0,T] \times E$. The function $f_0(\rho,u,t)$ is measurable, $f(\rho,u,t)$ is strongly measurable in t for ρ fixed and continuous in ρ for t fixed. For every $c > 0$ there exists $K_0(t,c)$, $L_0(t,c) \in L^1(0,T)$ such that

$$f_0(\rho,u,t) \leq K_0(t,c), \quad (0 \leq t \leq T, y \leq c) \tag{D.11}$$

$$\partial_\rho f_0(\rho,u,t) \leq L_0(t,c), \quad (0 \leq t \leq T, y \leq c) \tag{D.12}$$

$$f(\rho,u,t) \leq K(t,c), \quad (0 \leq t \leq T, y \leq c) \tag{D.13}$$

$$\partial_\rho f(\rho,u,t) \leq L(t,c), \quad (0 \leq t \leq T, y \leq c) \tag{D.14}$$

Finally, we assume that the Fréchet derivative $\partial_\rho g_0(\rho,u,t)$ exists in $[0,T] \times E$. All this properties do not depend on the control u which is in the set of admissible control.

Theorem D.1 [21]. (*The Minimum Principle for general control problems.*) Let us assume that the target set Y_ρ is closed and that $C_{ad}(0,T;U)$, the set of admissible control in the time interval $[0,T]$ taking values from set U, is patch complete and saturated. Then, if u^* is an optimal control in $0 \leq t \leq T$ there exists $(z_0,z_1) \in R \times E$, $z_0 \geq 0$, $z \in N_{Y_\rho}(\rho^*) = $ normal cone to Y_ρ at $\rho(T,u^*)$ such that, if $\pi(s)$ is the solution of

$$\partial_t \pi(s) = -[A(s) + \partial_\rho f(\rho(u,s),u(s),s)]^T \pi(s) \tag{D.15}$$
$$- z_0 \partial_\rho f_0(\rho(u^*,s),u^*(s),s)$$
$$\pi(t) = z + z_0 \partial_\rho g_0(\rho(u^*,T),T) \tag{D.16}$$

in $0 \leq s \leq T$

$$0 \leq z_0 \int_0^T \{[f_0(\rho(u^*,\sigma),v(\sigma),\sigma) - f_0(\rho(\sigma,u^*),u^*(\sigma),\sigma)]$$
$$+ \langle \pi(\sigma), f(\rho(\sigma,u^*),v(\sigma),\sigma) - f(\rho(\sigma,u^*),u^*(\sigma),\sigma) \rangle \} \, d\sigma$$
$$\tag{D.17}$$

for all $v \in C_{ad}(0,T;U)$. If E is separable, for $f_0(\rho(t,u^*),v(t),t)$ and $f(\rho(t,u^*),v(t),t)$ are regular in U and $C_{ad}(0,T;U)$ is spike complete, then

$$z_0 f_0(\rho(u^*,s),u^*(s),s) + \langle \pi(s), f(\rho(u^*,s),u^*(s),s) \rangle =$$
$$\min_{v \in U} z_0 f_0(\rho(u^*,s),v,s) + \langle \pi(s), f(\rho(s,u^*),v,s) \rangle \tag{D.18}$$

i.e., in $0 \leq s \leq T$

Remark. Introducing the Hamiltonian H:

$$H(\rho^*, v, t) = z_0 f_0(\rho(u^*, t), v, t) + \langle \pi(t), f(\rho(u^*, t), v, t) \rangle \tag{D.19}$$

expressions (D.17) and (D.18) can be written, respectively, as:

$$u^* = \min_{v \in C_{ad}} \int_0^T H(\rho^*, v, \sigma) d\sigma \tag{D.20}$$

$$u^*(t) = \min_{v(t) \in U} H(\sigma, \rho^*, v(t)) d\sigma \tag{D.21}$$

References

1. W. AGASSOUNON, A. MARTINOLI, AND K. EASTON, *Macroscopic Modeling of Aggregation Experiments using Embodied Agents in Teams of Constant and Time-Varying Sizes*, Autonomous Robots, 17 (2004), pp. 163–192.
2. R. ALBERT AND A. BARABASI, *Statistical Mechanics of Complex Networks*, Reviews of Modern Physics, 74 (2002), pp. 47–97.
3. R. ALUR, C. BELTA, F. IVANCIC, V. KUMAR, M. MINTZ, G. J. PAPPAS, H. RUBIN, AND J. SCHUG, *Hybrid Modeling and Simulation of Biomolecular Networks*, Hybrid Systems: Computation and Control, Lecture Notes in Computer Science, 2034 (2001), p. 19.
4. R. ARKIN, *Behavior-Based Robotics*, The MIT Press, 1998.
5. M. F. BACHMANN, M. SALZMANN, A. OXENIUS, AND P. S. OHASHI, *Formation of TCR dimers/trimers as a crucial step for T cell activation*, Eur. J. Immunol., 28 (1998), pp. 2571–2579.
6. T. BALCH AND L. PARKER, *Robot Teams: From Diversity to Polymorphism*, AK Peters, 2002.
7. C. BERGMANN, J. L. VAN HEMMEN, AND L. A. SEGEL, *Th1 or Th2: how an appropriate T helper response can be Made*, Bull. Math. Biol., 63 (2001), pp. 405–430.
8. D. BERTSEKAS, *Nonlinear Programming*, Athena Scientific, 1999.
9. G. BOCHAROV ET. AL., *Underwhelming the Immune Response: Effect of Slow Virus Growth on $CD8^+$-T-Lymphocyte Responses*, Journal of Virology, 78 (2004), pp. 2247–2254.
10. E. BONABEAU, G. THERAULAZ, AND J. DENEUBOURG, *Group and Mass Recruitment in Ant Colonies: the Influence of Contact Rates*, J. Theor. Biol., 195 (1998), pp. 157–166.
11. E. BONABEAU, G. THERAULAZ, J.-L. DENEUBOURG, S. ARON, AND S. CAMAZINE, *Self-Organization in social insects*, Trends Ecol. Evol., 12 (1997), pp. 188–193.
12. J. CARNEIRO, J. STEWART, A. COUTINHO, G. COUTINHO, AND C. MARTINEZ-A, *The ontogeny of class-regulation of CD4+ T lymphocyte populations*, Int. Immunol, 7 (1995), pp. 1265–1277.
13. A. CHAKRABORTY, *Lighting Up TCR Takes Advantage of Serial Triggering*, Nature Immunology, 3 (2002), pp. 895–896.
14. D. COOMBS, M. DEMBO, C. WOFSY, AND B. GOLDSTEIN, *Equilibrium thermodynamics of cell-cell adhesion mediated by multiple ligand-receptor pairs*, Biophysical Journal, 86 (2004), pp. 1408–1423.

15. T. COVER AND J. THOMAS, *Elements of Information Theory*, Wiley-Interscience, New York, 1991.

16. R. J. DE BOER, L. A. SEGEL, AND A. S. PERELSON, *Pattern formation in one- and two-dimensional shape-space models of the immune system*, J. Theor. Biol., 155 (1992), pp. 295–333.

17. C. DETRAIN, J. L. DENEUBOURG, AND J. M. PASTEELS, *Information Processing in Social Insects*, Birkhauser, 1999.

18. Z. DJUROVIĆ AND B. KOVAČEVIĆ, *Bandwidth selection for kernel density estimation based on QQ-plot*, WSEAS Trans. on Circuits and Systems, 3 (2004), pp. 679–687.

19. Z. DJUROVIĆ, B. KOVAČEVIĆ, AND V. BARROSO, *QQ-plot Based Probability Density Function Estimation*, Proc. 10-th Workshop Stat. Sig. and Array Processing, Pocono Manor, PA, USA, (2000).

20. L. C. EVANS, *Partial Differential Equations*, American Mathematical Society, Providence, 1998.

21. H. O. FATTORINI, *Infinite Dimensional Optimization and Control Theory*, Cambridge University Press, 1999.

22. C. FELLBAUM, *WordNet: An Electronic Lexical Database*, Bradford Books, 1998.

23. J. FERBER, *Multi-Agent Systems: An Introduction to Distributed Artificial Intelligence* , Addison-Wesley Longman, 1999.

24. D. FLAMM, *History and Outlook of Statistical Physics*, ArXiv Physics e-prints, http://www.citebase.org/abstract?id=oai:arXiv.org:physics/9803005, 1998.

25. C. W. GARDINER, *Handbook of Stochastic Methods for Physics, Chemistry and the Natural Sciences*, Springer, second ed., 1985.

26. A. GELB, *Applied Optimal Estimation*, The MIT Press, 1974.

27. M. GUNZER AND A. SCHAFER ET AL., *Antigen presentation in extracellular matrix: Interactions of T cells with dendritic cells are dynamic, short lived, and sequential*, Immunity, 13 (2000), pp. 323–332.

28. S. HARSDORF AND R. REUTER, *Stable Deconvolution of Noisy Lidar Signals*, EARSeL eProceedings, 1 (2001), pp. 88–95.

29. A. HIRSH AND D. GORDON, *Distributed problem solving in social insects*, Annals of Mathematics and Artificial Intelligence, 31 (2001), pp. 199–221.

30. HTTP://WWW.TRAINING.SEER.CANCER.GOV, *funded by the U.S. National Cancer Institute's Surveillance, Epidemiology and End Results (SEER) Program*, via contract number N01-CN-67006, with Emory University, Atlanta SEER Cancer Registry, Atlanta, Georgia, U.S.A.

31. J. B. HUPPA AND M. M. DAVIS, *T-cell-antigen recognition and the immunological synapse*, Nat. Rev. Immunol., 3 (2003), pp. 973–983.

32. A. JADBABAIE, J. LIN, AND A. S. MORSE, *Coordination of Groups of Mobile Autonomous Agents Using Nearest Neighbor Rules*, IEEE Trans. on Automatic Control, 48 (2003), pp. 988–1001.

33. C. A. JANEWAY AND P. TRAVERS, *Immunobiology*, Blackwell Scientific Publication, 1994.

34. E. T. JEYNES, *Information Theory and Statistical Mechanics*, Physical Review, 106 (1957), pp. 620–630.

35. H. JIANGHAI, J. LYGEROS, AND S. SASTRY, *Towards a Theory of Stochastic Hybrid Systems*, Hybrid Systems: Computation and Control, Lecture Notes in Computer Science, 1790 (2000), pp. 160–173.

36. M. KAUFMAN, F. ANDRIS, AND O. LEO, *A Logical Analysis of T Cell Activation and Anergy*, Immunology, Proc. Natl. Acad. Sci. USA, 96 (1999), pp. 3894–3899.

37. T. B. KEPLER AND A. S. PERELSON, *Somatic Hypermutation in B Cells: An Optimal Control Treatment*, J. Theor. Biol., 164 (1993), pp. 37–64.

38. H. KITANO, *Computational Systems Biology*, Nature, 420 (2002), pp. 206–210.

39. L. D. LANDAU AND E. M. LIFSHITZ, *Statistical Physics*, Pergamon Press, 1959.

40. K. H. LEE AND A. R. DINNER ET AL., *The Immunological Synapse Balances T Cell Receptor Signaling and Degradation*, Science, 302 (2003), pp. 1218–1222.

41. K. LÉON, A. LAGE, AND J. CARNEIRO, *Tolerance and immunity in a mathematical model of T-cell mediated suppression*, J. Theor. Biol., 255 (2003), pp. 107–126.

42. K. LÉON, R.PEREZ, A.LAGE, AND J.CARNEIRO, *Three-cell interactions in T cell mediated suppression? A mathematical analysis of its quantitative implications*, J. Immunol., 166 (2001), pp. 5356–5365.

43. K. LERMAN AND A. GALSTYAN, *Mathematical Model of Foraging in a Group of Robots: Effect of Interference*, Autonomous Robots, 13 (2002), pp. 127–141.

44. K. LERMAN, C. JONES, A. GALSTYAN, AND M. J. MATARIĆ, *Analysis of Dynamic Task Allocation in Multi-Robot Systems*, Int. J. of Robotics Research, 25 (2006), pp. 225–242.

45. A. LINO, *Caracterização do mecanismo molecular de activação de linfocitos T*, Relatório Final de Curso Bioqimica, Faculdade de Ciências da Univesidade de Lisboa, Lisboa pp.47, 2000.

46. J. L. LIONS, *Optimal Control of Systems Governed by Partial Differential Equations*, Springer-Verlag, 1971.

47. L. B. LUCY, *An Iterative Technique for the Rectification of Observed Distributions*, Astronomical Journal, 79 (1974), pp. 745–754.

48. A. MARTINOLI, K. EASTON, AND W. AGASSOUNON, *Modeling Swarm Robotic Systems: A Case Study in Collaborative Distributed Manipulation*, Int. Journal of Robotics Research, 23 (2004), pp. 415–436.

49. A. MARTINOLI, A. J. IJSPEERT, AND L. M. GAMBARDELLA, *A Probabilistic Model for Understanding and Comparing Collective Aggregation Mechanisms*, Proceedings of the 5th European Conference on Advances in Artificial Life (ECAL-99), volume of LNAI, 1674 (1999), pp. 575–584.

50. M.ATHANS AND P. FALB, *Optimal Control: An Introduction to the Theory and its Applications*, McGraw Hill, New York, 1966.

51. T. R. MEMPEL, S. E. HENRICKSON, AND U. H. VON ANDRIAN, *T cell priming by dendritic cells in lymph nodes occurs in three distinct phases*, Nature, 427 (2004), pp. 154–159.

52. M. J. MILLER, S. H. WEI, I. PARKER, AND M. D. CAHALAN, *Two-photon imaging of lymphocyte motility and antigen response in intact lymph node*, Science, (2002), pp. 1869–1873.

53. D. MILUTINOVIĆ, J. CARNEIRO, M. ATHANS, AND P. LIMA, *A Hybrid Automata Modell of TCR Triggering Dynamics*, Proceedings of the 11th IEEE Mediterranean Conference on Control and Automation (MED2003), Rhodes, Greece, (2003).

54. D. MILUTINOVIĆ AND J. CARNEIRO ET AL., *Modeling Dynamics of Cell Population Molecule Expression Distribution*, accepted for publishing by Nonlinear Analysis: Hybrid Systems, Elsevier, (2007).

55. D. MILUTINOVIĆ AND P. LIMA, *Modeling and Optimal Centralized Control of a Large-Size Robotic Population*, IEEE Transactions on Robotics, 22 (2006), pp. 1280–1285.

56. D. MILUTINOVIĆ, P. LIMA, AND M. ATHANS, *Biologically Inspired Stochastic Hybrid Control of Multi-Robot Systems*, Proceedings of the 11th International Conference on Advanced Robotics (ICAR), University of Coimbra, Portugal, (2003).

57. M. MORARI, *Hybrid system analysis and control via mixed integer optimization*, IFAC Symposium on Dynamics and Control of Process Systems, 1 (2002), pp. 1–12.

58. A. U. NEUMANN AND G. WEISBUCH, *Dynamics and topology of idiotypic networks*, Bull. Math. Biol., 54 (1992), pp. 699–726.

59. G. NICOLIS AND I. PRIGOGINE, *Self-Organization in Nonequilibrium Systems*, John Wiley and Sons Inc., 1977.

60. M. A. NOWAK AND R. M. MAY, *Virus Dynamics: Mathematical Principles of Immunology and Virology*, Oxford University Press, 2000.

61. A. S. PERELSON AND P. W. NELSON, *Mathematical Analysis of HIV-1 Dynamics in Vivo*, SIAM Review, 41 (1999), pp. 3–44.

62. A. S. PERELSON AND G. WEISBUCH, *Immunology for Physicists*, Reviews of Modern Physics, 69 (1997), pp. 1219–1268.

63. B. POLIC, D. KUNKEL, A. SCHEFFOLD, AND K. RAJEWSKY, *How alpha beta T cells deal with induced TCR alpha ablation*, Proc. Natl. Acad. Sci. USA, 98 (2001), pp. 8744–8749.

64. M. J. D. POWELL, *Restart procedures for the conjugate gradient method*, Mathematical Programming, 12 (1977), pp. 241–254.

65. C. W. REYNOLDS, *Flocks, herds and schools: A distributed behavioral model*, Computer Graphics, 21 (1987), pp. 25–34.

66. W. H. RICHARDSON, *Bayesian-Based Iterative Method of Image Restoration*, Journal of the Optical Society of America, 62 (1972), pp. 55–59.

67. L. SEGEL AND A. PERELSON, *Shape space: An Approach to the Evaluation of Cross-Reactivity Effects, Stability and Controllability in the Immune System*, Immunol Lett., 22 (1989), pp. 91–99.

68. J. SOUSA, *Modeling the antigen and cytokine receptors signaling processes and their propagation to lymphocyte population dynamics*, PhD Dissertation, University of Lisbon, Portugal, 2003.

69. J. SOUSA AND J. CARNEIRO, *A Mathematical Analysis of TCR Serial Triggering and Down-Regulation*, Eur. J. Immunol., 30 (2000), pp. 3219–3227.

70. J. STEWART AND F. J. VARELA, *Morphogenesis in shape-space. Elementary meta-dynamics in a model of the immune network*, J. Theor. Biol., 153 (1991), pp. 477–498.

71. K. SUGAWARA AND M. SANO, *Cooperative acceleration of task performance: Foraging behavior of interacting multi-robot system*, Physica D, 100 (1997), pp. 343–354.

72. B. SUZER, J. L. VAN HEMMEN, A. U. NEUMANN, AND U. BEHN, *Memory in idiotypic networks due to competition between proliferation and differentiation*, Bull. Math. Biol., 55 (1993), pp. 1133–1182.

73. P. TABUADA, G. J. PAPPAS, AND P. LIMA, *Feasible formations of multi-agent systems*, Proc. American Control Conf. Arlington, 1 (2001), pp. 56–61.

74. H. G. TANNER, G. J. PAPPAS, AND V. KUMAR, *Leader-to-Formation Stability*, IEEE Trans. on Robotics and Automation, 20 (2004), pp. 443–455.

75. S. VALITUTTI, S. MULLER, M. CELLA, E. PADOVAN, AND A. LANZAVECCHIA, *Serial Triggering of Many T-cell Receptors by a Few Peptide-MHC Complexes*, Nature, 375 (1995), pp. 148–151.

76. A. J. VAN DER SCHAFT AND J. M. SCHUMACHER, *An Introduction to Hybrid Dynamical Systems*, Springer-Verlag, 2000.

77. H. VON BOEHMER AND I. AIFANTIS ET AL., *Thymic selection revisited: how essential is it?*, Immunol. Rev., 191 (2003), pp. 62–78.

78. O. VON STRYK AND R. BULIRSCH, *Direct and Indirect Methods for Trajectory Optimization*, Annals of Operations Research, 37 (1992), pp. 357–373.

79. L. M. WEIN, S. A. ZENIOS, AND M. A. NOWAK, *Dynamic Multidrug Therapies for HIV: A Control Theoretic Approach*, J. Theor. Biol., 185 (1997), pp. 15–29.

80. G. WEISBUCH, R. J. DE BOER, AND A. S. PERELSON, *Localized memories in idiotypic networks*, J. Theor. Biol., 146 (1990), pp. 483–499.

81. D. WITHERDEN AND N. VAN OERS ET AL., *Tetracycline-controllable selection of CD4(+) T cells: half-life and survival signals in the absence of major histocompatibility complex class II molecules*, J. Exp. Med., 191 (2000), pp. 355–364.

82. D. WODARZ, K. M. PAGE, R. A. ARNAOUT, A. R. THOMSEN, J. D. LIFSON, AND M. A. NOWAK, *A New Theory of Cytotoxic T-lymphocyte Memory: Implication for HIV Treatment*, Philos. Trans. Royal Soc. Lond. B Biol. Sci., 355 (2000), pp. 329–343.

83. C. WOFSY AND B. GOLDSTEIN, *Effective Rate Models for Receptors Distributed in a Layer above a Surface: Application to Cells and Biacore*, Biophysical Journal, 82 (2002), pp. 1743–1755.

84. M. WOOLDRIDGE, *An Introduction to MultiAgent Systems*, John Wiley and Sons, 2002.

85. O. C. ZIENKIEWICZ AND R. L. TAYLOR, *The Finite Element Method - Fluid Dynamics*, vol. 3, Butterworth-Heinemann, 2000.

Index

Springer Tracts in Advanced Robotics

Edited by B. Siciliano, O. Khatib and F. Groen

Vol. 32: Milutinović, D.; Lima, P.
Cells and Robots: Modeling and Control of
Large-Size Agent Populations
124 p. 2007 [978-3-540-71981-6]

Vol. 31: Ferre, M.; Buss, M.; Aracil, R.;
Melchiorri, C.; Balaguer C. (Eds.)
Advances in Telerobotics
500 p. 2007 [978-3-540-71363-0]

Vol. 30: Brugali, D. (Ed.)
Software Engineering for Experimental Robotics
490 p. 2007 [978-3-540-68949-2]

Vol. 29: Secchi, C.; Stramigioli, S.; Fantuzzi, C.
Control of Interactive Robotic Interfaces – A
Port-Hamiltonian Approach
225 p. 2007 [978-3-540-49712-7]

Vol. 28: Thrun, S.; Brooks, R.; Durrant-Whyte, H.
(Eds.)
Robotics Research – Results of the 12th
International Symposium ISRR
602 p. 2007 [978-3-540-48110-2]

Vol. 27: Montemerlo, M.; Thrun, S.
FastSLAM – A Scalable Method for the
Simultaneous Localization and Mapping
Problem in Robotics
120 p. 2007 [978-3-540-46399-3]

Vol. 26: Taylor, G.; Kleeman, L.
Visual Perception and Robotic Manipulation – 3D
Object Recognition, Tracking and Hand-Eye
Coordination
218 p. 2007 [978-3-540-33454-5]

Vol. 25: Corke, P.; Sukkarieh, S. (Eds.)
Field and Service Robotics – Results of the 5th
International Conference
580 p. 2006 [978-3-540-33452-1]

Vol. 24: Yuta, S.; Asama, H.; Thrun, S.;
Prassler, E.; Tsubouchi, T. (Eds.)
Field and Service Robotics – Recent Advances in
Research and Applications
550 p. 2006 [978-3-540-32801-8]

Vol. 23: Andrade-Cetto, J.; Sanfeliu, A.
Environment Learning for Indoor Mobile Robots
– A Stochastic State Estimation Approach
to Simultaneous Localization and Map Building
130 p. 2006 [978-3-540-32795-0]

Vol. 22: Christensen, H.I. (Ed.)
European Robotics Symposium 2006
209 p. 2006 [978-3-540-32688-5]

Vol. 21: Ang Jr., H.; Khatib, O. (Eds.)
Experimental Robotics IX – The 9th International
Symposium on Experimental Robotics
618 p. 2006 [978-3-540-28816-9]

Vol. 20: Xu, Y.; Ou, Y.
Control of Single Wheel Robots
188 p. 2005 [978-3-540-28184-9]

Vol. 19: Lefebvre, T.; Bruyninckx, H.;
De Schutter, J. Nonlinear Kalman Filtering
for Force-Controlled Robot Tasks
280 p. 2005 [978-3-540-28023-1]

Vol. 18: Barbagli, F.; Prattichizzo, D.;
Salisbury, K. (Eds.)
Multi-point Interaction with Real
and Virtual Objects
281 p. 2005 [978-3-540-26036-3]

Vol. 17: Erdmann, M.; Hsu, D.; Overmars, M.;
van der Stappen, F.A (Eds.)
Algorithmic Foundations of Robotics VI
472 p. 2005 [978-3-540-25728-8]

Vol. 16: Cuesta, F.; Ollero, A.
Intelligent Mobile Robot Navigation
224 p. 2005 [978-3-540-23956-7]

Vol. 15: Dario, P.; Chatila R. (Eds.)
Robotics Research – The Eleventh
International Symposium
595 p. 2005 [978-3-540-23214-8]

Vol. 14: Prassler, E.; Lawitzky, G.; Stopp, A.;
Grunwald, G.; Hägele, M.; Dillmann, R.;
Iossifidis. I. (Eds.)
Advances in Human-Robot Interaction
414 p. 2005 [978-3-540-23211-7]

Vol. 13: Chung, W.
Nonholonomic Manipulators
115 p. 2004 [978-3-540-22108-1]

Vol. 12: Iagnemma K.; Dubowsky, S.
Mobile Robots in Rough Terrain –
Estimation, Motion Planning, and Control
with Application to Planetary Rovers
123 p. 2004 [978-3-540-21968-2]

Vol. 11: Kim, J.-H.; Kim, D.-H.; Kim, Y.-J.;
Seow, K.-T.
Soccer Robotics
353 p. 2004 [978-3-540-21859-3]

Vol. 10: Siciliano, B.; De Luca, A.; Melchiorri, C.;
Casalino, G. (Eds.)
Advances in Control of Articulated and
Mobile Robots
259 p. 2004 [978-3-540-20783-2]

Vol. 9: Yamane, K.
Simulating and Generating Motions
of Human Figures
176 p. 2004 [978-3-540-20317-9]

Vol. 8: Baeten, J.; De Schutter, J.
Integrated Visual Servoing and Force
Control – The Task Frame Approach
198 p. 2004 [978-3-540-40475-0]

Vol. 7: Boissonnat, J.-D.; Burdick, J.;
Goldberg, K.; Hutchinson, S. (Eds.)
Algorithmic Foundations of Robotics V
577 p. 2004 [978-3-540-40476-7]

Vol. 6: Jarvis, R.A.; Zelinsky, A. (Eds.)
Robotics Research – The Tenth
International Symposium
580 p. 2003 [978-3-540-00550-6]

Vol. 5: Siciliano, B.; Dario, P. (Eds.)
Experimental Robotics VIII – Proceedings
of the 8th International Symposium ISER02
685 p. 2003 [978-3-540-00305-2]

Vol. 4: Bicchi, A.; Christensen, H.I.;
Prattichizzo, D. (Eds.)
Control Problems in Robotics
296 p. 2003 [978-3-540-00251-2]

Vol. 3: Natale, C.
Interaction Control of Robot Manipulators –
Six-degrees-of-freedom Tasks
120 p. 2003 [978-3-540-00159-1]

Vol. 2: Antonelli, G.
Underwater Robots – Motion and Force Control
of Vehicle-Manipulator Systems
268 p. 2006 [978-3-540-31752-4]

Vol. 1: Caccavale, F.; Villani, L. (Eds.)
Fault Diagnosis and Fault Tolerance for
Mechatronic Systems – Recent Advances
191 p. 2003 [978-3-540-44159-5]

®
FSC
www.fsc.org
MIX
Papier aus verantwortungsvollen Quellen
Paper from responsible sources
FSC® C105338